Acoustics of Multi-Use Performing Arts Centers

Acoustics of Multi-Use Performing Arts Centers

Mark Holden

CRC Press
Taylor & Francis Group
Boca Raton London New York

CRC Press is an imprint of the
Taylor & Francis Group, an **informa** business

A SPON PRESS BOOK

CRC Press
Taylor & Francis Group
6000 Broken Sound Parkway NW, Suite 300
Boca Raton, FL 33487-2742

First issued in paperback 2019

© 2016 by Mark Ashton Holden
CRC Press is an imprint of Taylor & Francis Group, an Informa business

No claim to original U.S. Government works

ISBN-13: 978-0-415-51719-5 (hbk)
ISBN-13: 978-0-367-86610-5 (pbk)

Visit the Taylor & Francis Web site at
http://www.taylorandfrancis.com

and the CRC Press Web site at
http://www.crcpress.com

Contents

SECTION II CREATING THE BUILDING

SECTION III   ARCHITECTURAL DETAILS

SECTION IV MEASURING RESULTS

SECTION V CASE STUDIES

Acknowledgments

This book would not have been possible without the generous and dedicated assistance of many. I fear that I might leave someone out but thank everyone who contributed to getting this modest work completed. When Taylor & Francis approached me with the idea to write a text on acoustics within multi-use halls, I agreed because I believed that I had something unique to contribute.

I began this book with the naïve idea that it could be completed in a couple of years, in the kitchen, on Saturday and Sunday mornings. My optimism stretched for another year and a half, and then I compelled the team at Jaffe Holden and staff at Malcolm Holzman's office to help me complete the editing and illustrations. For their assistance, I am eternally grateful.

Much of what I have expressed here was learned from many expert acousticians and scientists who paved the way with their research, testing, calculations, and measurements. I have acknowledged specific instances where I have used others' work, but much of it was learned in technical meetings of the Acoustical Society of America and Audio Engineering Society (AES) and in conversations with great acousticians such as my mentor, the late J. Christopher Jaffe, and with Leo Beranek, the father of everything known about acoustics.

Specifically, I am in awe of Jaffe Holden's director of marketing and communications, Paige Williams, and her tireless reworking of nearly every convoluted sentence I wrote. She kept me organized and on task, gave me encouragement when I needed it, and task mastered me when I did not (well, really I did). Paige, I owe you so much. Thank you.

The director of operations at Jaffe Holden, Sig Hauck, never lost confidence in me or this project and cleared one roadblock after another to get it completed even when I ran well over budget and schedule! He tolerated my distractedness from my rainmaking efforts and design duties and never lost confidence that this would be completed. Sig, I thank you!

Many thanks to Mathew Rosenthal for completing many technical charts and graphs, Steven Schlaseman for his expert authorship on the glossary and technical review, and Russ Cooper for his kind review and insights. Mark Turpin provided insight for the electronic architecture portion of Chapter 17, Carlos Rivera located hard-to-find case study data, and Matthew Nichols took expert photography of samples, the smoke effect, and even the cover shot.

I am indebted to Malcolm Holzman and Kurt Wehmann (as well as Rafael Ayala, Ermira Kasapi, and Won Kim) of Holzman Moss Bottino Architecture, based in New York, who stepped up to draw (and redraw) more than 100 illustrations and drawings that are so critical to understanding the concepts and ideas of the text. Malcolm, author of a number of excellent architectural texts, uniquely understood my challenge. Many, many thanks to all of you.

Thank you to the following companies and individuals who provided support in various forms including data, information, and access for photography:

■ Lincoln Center for the Performing Arts in New York City
■ Theatre Projects Consultants
■ Shen Milsom & Wilke
■ Wilson Butler Architects
■ Steve Barber

Finally, I want to thank my family—my sons John and Luke and, especially, my dear wife Becky—for their patience, tolerance, and understanding of my obsession with writing this book and why it was important for me to share some of what I have learned over the last 35 years.

Mark Holden

Author

Mark Holden has explored acoustics within orchestra shells, orchestra pits, attics, and audience chambers in multi-use performing arts centers around the globe. Many of Mark's passions are complemented by his work as an acoustician, including classical and jazz music, the physics of wave energy, mathematics, and the design of highly complex buildings. His mission is to create aural environments that transport audiences and performers to new aesthetic and spiritual levels.

Mark was born into a family of musicians and scientists. His great grandfather Hubert Holden participated in the creation of the first direct drive motorcycle in 1914, the development of the Brooklands motor racing track with banked curves, and the development of an improved World War II anti-aircraft gun. Mark's mother, Jean Fisher, is an accomplished soprano who auditioned for the West End premiere of *My Fair Lady* but lost the part to the venerable Julie Andrews. Jean sang for years with the Cleveland Orchestra Chorus at Blossom Music Center, a Chris Jaffe–designed outdoor pavilion.

Born in Toronto, Canada, and educated at Western Reserve Academy prep school in Hudson, Ohio, Mark attended Duke University and earned a bachelor of science degree in 1978 in electrical engineering. He participated in the jazz ensemble and performed in Baldwin Hall in 1977. During his time at Duke, Mark became enamored with acoustics while participating in an acoustic study of Duke's student union as part of a physics course. This experience came full circle when Mark worked on the renovation of Duke's Baldwin Hall (explored as a case study within this book) and further services on a new student union building at Duke.

Mark has authored columns for major trade publications and lectured at Harvard's School of Design and at Duke's Pratt School of Engineering. Mark is a member of the National Council of Acoustical Consultants, a member of the United States Institute of Theater Technology (USITT), and an elected fellow of the Acoustical Society of America. He regularly presents at conferences for USITT and the American Institute of Architects.

Mark lives in Connecticut with his wife, Becky, an accomplished visual artist and educator, and two adult sons, John and Luke. When not traveling, he enjoys jogging, creating dishes in his carefully designed kitchen, and tending to his extensive collection of perennial flower gardens.

Introduction

Acoustics of Multi-Use Performing Arts Centers

Multi-use performing arts centers were once considered pariahs of the arts community. Through the use of adjustable acoustics systems, these types of halls can now adapt to different types of performance without degradation in sound quality and are comparable to many concert halls and single-purpose halls. My passion for the complexity and artistry required in the acoustic design of these spaces is revealed within this book. I hope that it serves to enrich the reader.

This book is a step-by-step manual on how to achieve outstanding acoustics in multi-use performance spaces. I will guide the reader from planning of the initial concept to the final tuning. This book is a tool for architects, acousticians, musicians, and students in addition to the general public. It is important to note that this book is informed by evidence-based design gleaned from real-world experience and not just theory. Only necessary mathematics and terminology explanations are included within this book. A glossary includes more in-depth definitions and derivations.

This book is structured into the following sections:

- Building Blocks: Chapters 1–4
    - This section covers the fundamentals of acoustics as it relates to initial stages of multi-use hall design in order to provide a solid foundation for the reader.
- Creating the Building: Chapters 5–8
    - In this section, concepts of acoustics are explored in terms of new and renovated spaces, and the basic components of the building structure are defined.
- Architectural Details: Chapters 9–14
    - This section examines floors, walls, ceilings, shells, and finishes and how they can be designed to achieve acoustic excellence.
- Measuring Results: Chapters 15–17
    - This section discusses how to use and tune adjustable acoustic systems in a multi-use hall in order to achieve acoustic excellence.
- Case Studies: Appendix
    - A collection of case studies on both new and renovated facilities is included in this section to demonstrate successful acoustic attributes and design.

It has been an honor and a privilege to work with talented architects, project managers owners, engineers, contractors and artists on multi-use performing arts centers around the world. I have collaborated with and learned from nine Pritzker Architecture Prize winners and am truly grateful for the many teachers I have had over the past 38 years. There is still much to learn about the science of acoustics and how the ear and brain interact. I look forward to continuing this journey and sharing it with my readers.

Mark Holden

BUILDING BLOCKS

I

Chapter 1

Making the Case for a Multi-Use Hall

Introduction

Purpose-built halls that serve solely as concert halls or opera houses are increasingly rare today because of high construction and operational costs. Only a few major international cities and high-profile institutions with deep pockets can afford them. Single-purpose halls have the advantage of being able to provide an ideal acoustic, theatrical, and artistic environment for each art form in individual facilities. The symphony can rehearse on the stage unencumbered by other performers needing the facility. The opera only needs to share its home for occasional outside performances. The theater can arrange sets that remain in place for extended periods of time (see Figure 1.1).

This exclusivity comes at significant capital and operational cost. In the United States, there is pressure by civic and business leaders for halls to consistently attract large audiences who pay to park, dine, and shop.

Capital costs for single-purpose facilities are substantial. For example, Kansas City, Missouri, privately raised more than $400 million for separate ballet/opera and symphony halls, and over a quarter of that sum was donated by a single foundation. The Kauffman Center for the Performing Arts opened in 2011 and features an 1800-seat ballet/opera house and a 1600-seat pure concert hall (see Figure 1.2).

The New World Center, designed by Frank Gehry, is a pure concert hall in Miami Beach that opened in 2011. This building cost $160 million and features innovative video display systems, excellent acoustics, and high-tech communication systems. The stage is nearly as large as the hall's seating area and is a viable financial model only because unique teaching and presenting opportunities exist for the space. Academic institutions with endowments, tuition, and donors can indeed build and operate intimate purpose-built halls for use by students and faculty. Jaffe Holden has collaborated on dozens of successful models like the New World Symphony Hall. However, it would be a mistake to assume that this hall design is the rule when, in fact, it is the exception.

Figure 1.1   Carnegie Hall, New York, NY, 1891. This iconic, purpose-built concert hall is known for excellent symphonic acoustics but is not well suited for opera, dance, or theatrical productions.

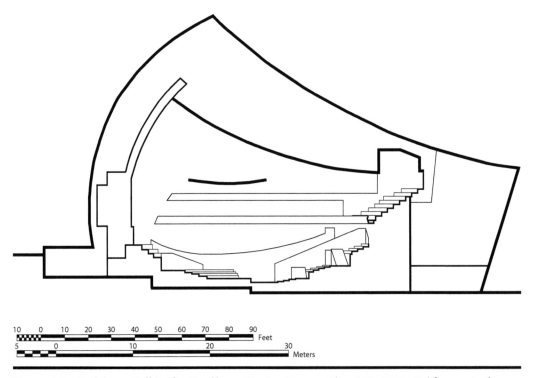

Figure 1.2   Helzberg Hall in the Kauffman Center, Kansas City, MO, 2011. With a cost of more than $400 million, this 1600-seat hall is an excellent concert hall but is less suited for dance, theater, and opera productions.

Short History

Despite the recent increase in popularity, the multi-use hall is not a new invention. The use of this type of building dates back to the 1920s. Although many aspects have changed over the years, the reasoning behind implementing this design has remained largely the same.

The 1920s and 1930s

Grand but technically unsuccessful municipal auditoriums in American cities during the 1920s and 1930s paved the way for the emergence of the multi-use hall. The municipal auditorium was the result of pressure from the artistic community for a large performance facility that would further the artistic development of local symphony, opera, and theater companies, as well as serve as a convention hall, grand ballroom, and ceremonial space (see Figure 1.3).

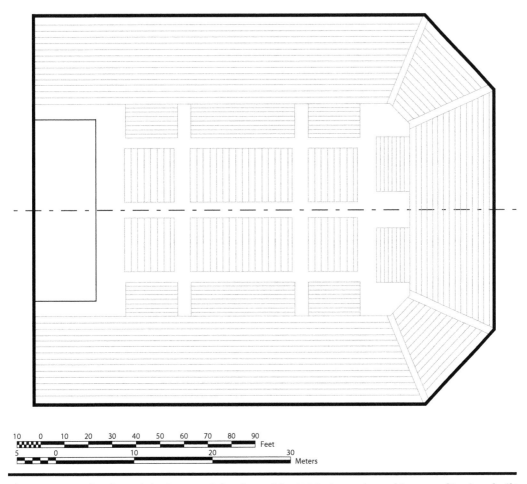

Figure 1.3 Columbus Civic Center, Columbus, GA, 1926. An early multi-use auditorium built to house performing arts, sports, exhibits, conferences, and political conventions. Acoustics are poor for all functions.

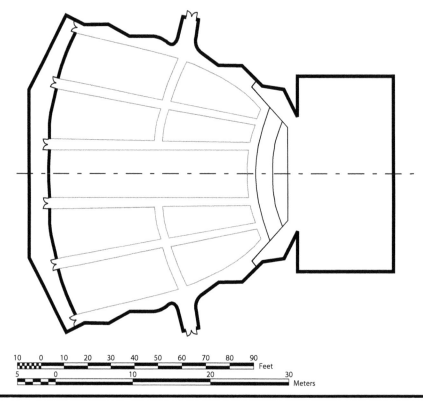

Figure 1.4 Jacksonville Civic Auditorium, Jacksonville, FL, 1962. This wide, 2000-seat fan-shaped multi-use hall was an improvement over earlier civic centers but had poor acoustics and was gutted in 1990.

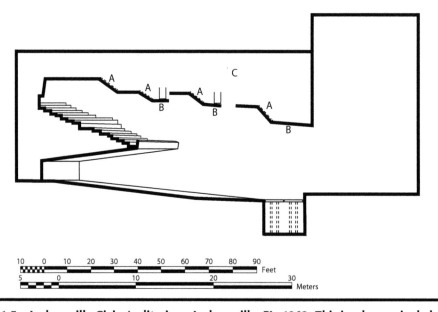

Figure 1.5 Jacksonville Civic Auditorium, Jacksonville, FL, 1962. This is a large, single balcony space. (A) Coffin-shaped ceiling openings. (B) Ceiling reflectors. (C) Upper acoustic volume.

The 1940s

During this decade, manufacturing and building construction industries focused on war efforts. As a result, very few halls were built during the 1940s (see Figure 1.3).

The 1950s and 1960s

After World War II, there was a growing desire from the public for modern halls to take on a more egalitarian form and thus eliminate exclusive boxes and grand tiers that created barriers between audiences. As part of this postwar civic expansionism and pride, democratic civic buildings were designed as homes for symphonies, theater troupes, and community concerts. Some also served as a war memorial or provided office space. In the 1950s and 1960s, cities such as Austin, Charleston, and Memphis built what was then considered to be fantastic new facilities. At the time they were created, these single-balcony, wide fan-shaped halls were considered to be state of the art technically, acoustically, and artistically.

It is now known that these halls lacked sonic impact and intimacy for theater, provided limited presence and clarity for opera, and were devoid of warm, rich reverberation for symphony. Still, they were a huge improvement over the barn-like convention centers that had served the communities for prior decades (see Figures 1.4 and 1.5).

The 1970s and 1980s

A new breed of multi-use halls came about in the 1970s and 1980s that were a vast improvement visually and theatrically but not much better acoustically. These buildings were technologically quite sophisticated and often employed moving ceilings to close off hall volume and create adjustable acoustic environments that met reverberation requirements and reduced seat capacity. Counterweighted, multiton steel contrivances supported the ceilings, catwalks, and lights but provided only a gross level of acoustic tunability and variability. Similar multi-use halls with subpar acoustics can be found in Northern Alberta, Canada, and Tokyo, Japan. In all fairness, these halls utilized the best available acoustic knowledge and consultants, but the tools available at the time were crude and unwieldy (see Figures 1.6 through 1.9).

Tools of the Trade

Manufacturers had few tools other than winches, cables, and counterweights to offer acousticians throughout the 1980s. Frankly, adjustable acoustic devices were more closely related to rigging ship anchors than to the needs of orchestras and opera companies. It is not a surprise that the multi-use halls got a bad reputation within the musical community. They were no match for the well-known pure concert halls such as Symphony Hall in Boston and Carnegie Hall in New York.

The 1990s

In the 1990s, Jaffe Holden set about to solve the conundrum of providing acoustic excellence for symphonic performances while at the same time meeting the theatrical and acoustic needs of other types of performance. Three new design directions were developed based on a facility's needs, end users, and budget. The first involved a sophisticated orchestra shell called the concert hall shaper. The second employed a system of double pit lifts to bring the orchestra past the proscenium and

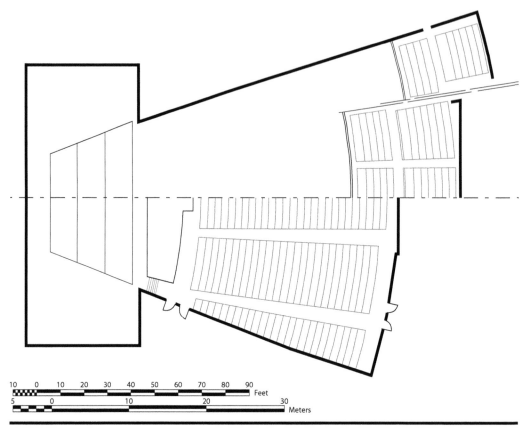

Figure 1.6 **Northern Alberta Jubilee Auditorium, Edmonton, AB, Canada, 1957. A 2500-seat fan-shaped hall with shallow balconies and low acoustic volume. Remodeled in 2004.**

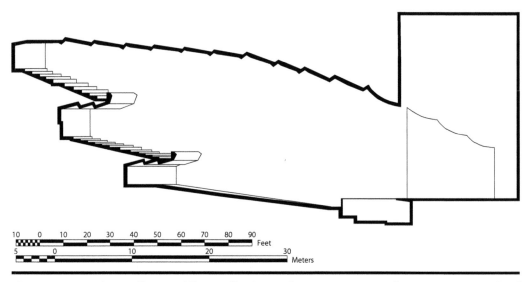

Figure 1.7 **Northern Alberta Jubilee Auditorium, Edmonton, AB, Canada, 1957. An example of an acoustically poor multi-use halls from the 1950s built in North America that tarnished the reputation of multi-use halls. Remodeled in 2004.**

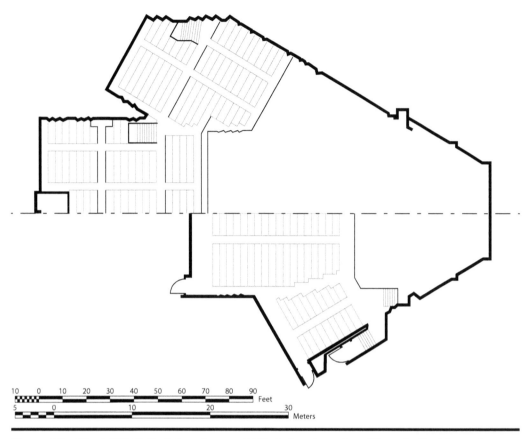

10 0 10 20 30 40 50 60 70 80 90
Feet
5 0 10 20 30
Meters

Figure 1.8 The NHK Hall, Tokyo, Japan, 1955. A fan-shaped multi-use hall with seating sections in terraces and low ceiling height.

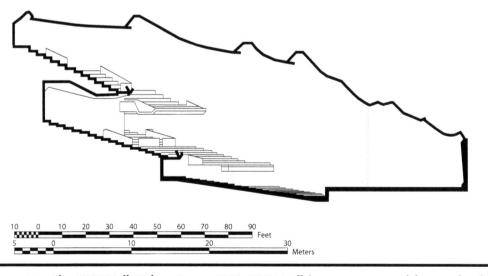

10 0 10 20 30 40 50 60 70 80 90
Feet
5 0 10 20 30
Meters

Figure 1.9 The NHK Hall, Tokyo, Japan, 1955. NHK Hall is a more successful example of a 1950's multi-use hall. Note that the wide fan shape is divided into smaller seating zones.

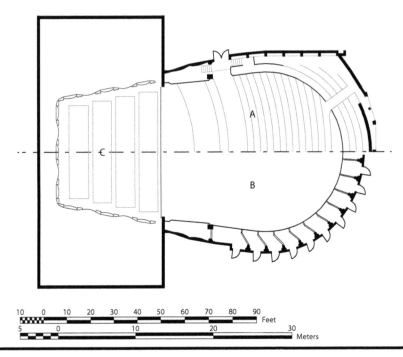

Figure 1.10 Bass Performance Hall, Fort Worth, TX, 1998. A 2000-seat multi-use hall with excellent acoustics, detailed in the Case Studies. (A) Orchestra level plan. (B) Box tier plan. (C) Concert Hall Shaper orchestra shell.

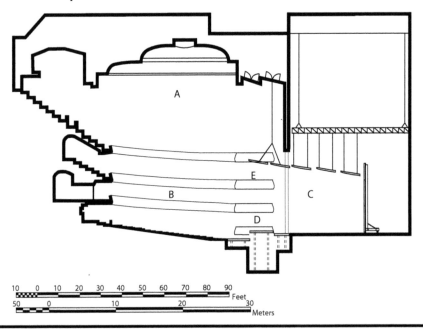

Figure 1.11 Bass Performance Hall, Fort Worth, TX, 1998. (A) High ceiling for proper acoustic volume with adjustable acoustic drapes. (B) Shallow balconies and side boxes for acoustic reflections and adjustable acoustic banners. (C) Tunable Concert Hall Shaper orchestra shell. (D) Orchestra pit/stage extension. (E) Forestage reflector.

out into the hall so that the auditorium could function as a one-room concert hall. The third approach placed a modified orchestra shell behind the proscenium and around the ensemble to project, blend, and aid onstage hearing.

Creating a flexible hall with excellent acoustics for classical music was but one piece of the acoustic conundrum. The acoustic challenges involving opera, amplified music, amplified musicals, and film or video presentations still needed to be addressed (see Figures 1.10 through 1.13).

Figure 1.12   Bass Performance Hall, Fort Worth, TX, 1998. Features complete adjustable acoustics technology within a classically designed hall. Featured as a case study.

Figure 1.13   Dell Hall at Long CPA, Austin, TX, 2008. A 2400-seat multi-use hall built within the shell of a 1950s civic auditorium. The green renovation showcases many of the adjustable acoustic systems described in this book at a modest budget. This hall is featured as a case study.

The Need for Multi-Use Halls

Communities that have a need for multi-use performance halls often hesitate to embrace them because of the misconception that they are not acoustically successful. Those quoting acoustic disasters in flexible halls have plenty of ammunition to draw from, as there have been more than a few spectacular failures. Acoustics as a "black art" is a common explanation for these failures. A careful study will reveal that poor acoustic results are often rooted in uninformed clients, bad design, team chemistry, inadequate funding, or inexperienced acoustic designers.

Shortfalls

Why is it that the acoustics of some halls have failed to meet expectations?

Lackluster Partnership

Acousticians have advanced computer modeling systems and years of field experience. Yet, there are halls that do not meet acoustic expectations. My theory is that, while the science of sound is quite well understood, the successful design of these halls requires more than raw scientific facts and years of theoretical experience. It requires a design team that works in total collaboration and partnership. Relationships must be carefully nurtured. I believe that relationships have a direct correlation to the outcome of the design and the acoustics of a space. A team that does not work well together and cannot communicate is likely to create a substandard project.

Pressure from Donors

While well intentioned, those that fund the construction costs of a new hall often have high goals and expectations well beyond acoustics. Donors know that sound is significant but place a great deal of importance on the ability of the new hall to stand as an icon especially since their name may be on the building. When the idea of the hall becomes more important to the donor than its actual functionality, decisions can often made that compromise acoustics.

Compromised Upgrades

Compromises to acoustic quality can sneak up on the design team. For example, well-intentioned designers or contractors could swap out materials without consulting the acousticians. Less-expensive wall materials often have a lower acoustic mass, seat upholstery perceived to be upgraded often absorbs more sound, and greener air systems are often noisier than the specs reviewed by the acoustician.

Communicating Ideas

The decisions and design solutions that make the difference between a successful hall and a mediocre one occur when the design and construction team work in total collaboration. When the team communicates and trusts each other, the hall is much more likely to succeed. Since every little detail can affect the acoustics of a hall, it is necessary for the acoustician to be involved in every design decision.

Multi-use hall design is often considered to be the most complex design. Contractors who have built nuclear power plants and military research labs agree that concert halls, theaters, and flexible performance halls are by far the most complex projects. Any slight change can affect the acoustic, structural, mechanical, theatrical, and code compliance in unintended ways. This domino effect is what renders multi-use hall design both complex and exciting.

Everyone in the business of hall design must have strong self-confidence in order to make their case during design meetings. However, an exaggerated ego can shut down the collaboration that is necessary for great buildings and great acoustics. It is important to be a good listener and a clear communicator.

Fear of losing control can also halt the open communication of ideas. The mentality of "my way or the highway" does not make for great collaboration. The stakes are very high. Reputations and prestige are on the line, as are millions of dollars. Holding on too tightly and micromanaging details can quickly throw a project off track just as trusting in the team's abilities encourages respect. In my experience, I have learned that respect and trust come more quickly when the team has a successful history of working on projects together. Good relationships strongly affect the outcome of any project. A team that does not work well together is doomed to create a substandard facility.

Success Stories

Engaging the acoustician very early in the design process results in successful halls. Ideally, acousticians begin the collaboration with donors, owners, and performing arts stakeholders when the initial design concepts are first put forth by the architects. In the best-case scenario, the acoustician is engaged before the rest of the design team and has a hand in influencing the selection of the design architect. In this scenario, it is assured that the lead designer will engage the acoustician in a meaningful way.

Individual groups acting independently, or within silos, often exist on a multidisciplinary design team. The success of the hall depends upon close collaboration and the removal of these silos. Acousticians; theater consultants; architects; interior designers; contractors; and structural, electrical, and mechanical engineers must form one cohesive and respectful team in order to achieve success.

There are a number of examples of multi-use halls that have excellent acoustics for various performance types, are affordable to build and operate, and revitalize communities. The Long Center for the Performing Arts in Austin, Texas, was a well-loved municipal auditorium built in the 1950s that has since been successfully transformed into two very flexible halls: the Michael & Susan Dell Hall and the Debra and Kevin Rollins Studio Theatre. These two spaces meet the needs of the Austin Symphony Orchestra, Austin Lyric Opera, and Ballet Austin in addition to touring Broadway productions, headliner acts, and local productions.

The case studies in the appendix of this book showcase successful halls, including Dell Hall, to more fully illustrate how a flexible-use hall can blossom into a magnificent facility for the community.

Chapter 2

The Building Block
of Reverberation

Introduction

What makes for excellent acoustics in some halls and what defines poor acoustics in others can be confusing to both the average listener and the sophisticated performer. Oftentimes, poor acoustics are blamed on the lack of wood materials in the walls or floor or on the excessive number of seats in a hall. Some myths for ensuring excellent acoustics are that concrete makes for bad acoustics, or placing broken wine bottles under the stage improves sound, as discussed by Goldsmith (2014)in *Discord: The Story of Noise*. These are both untrue and are proven as such by a number of reputable halls.

Both the Wiener Musikverein in Vienna, Austria, and the Concertgebouw in Amsterdam, Netherlands, are considered the gold standard for symphonic acoustics and have limited wood surface areas. The wood floor under the audience in the Concertgebouw is directly attached to a substantial concrete floor slab. The halls that achieve the highest acoustic ranking are built primarily of heavy plaster, which is more closely related to concrete in weight and chemical composition than wood. Some of these buildings have heavy wood timber under the plaster, but there also are examples of buildings structured with steel and concrete that have excellent symphonic acoustics, such as the Nancy Lee and Perry R. Bass Performance Hall in Fort Worth, Texas, or the Morton H. Myerson Symphony Center in Dallas, Texas.

Defining the Programming

The acoustics of a space must be determined by the particular type of performance that utilizes the space. Halls that are excellent for one performance type, amplified music for example, are often terrible for other types like classical symphonic performances. Radio City Music Hall in New York City demonstrates this point. The 6000-seat theater excels at presenting loud amplified popular music but fails at presenting operatic or unamplified symphonic music (see Figure 2.1).

Figure 2.1 Radio City Music Hall, New York, NY, 1932, 6000 seats, renovated by Jaffe Holden in 1999. (A) Sound absorptive rear wall. (B) Ceiling sound absorptive cast plaster now painted and sound reflective. Reverberation time 1.5 s, mid freq.

Figure 2.2 Overlay of Carnegie Hall section (in gray) with Radio City Music Hall section. Carnegie Hall is acoustically excellent for symphony concerts but not for highly amplified shows. Radio City Music Hall is acoustically excellent for amplified productions but not for the symphony.

Radio City Music Hall has a very large air volume, a concave ceiling, concave wall shapes, carpet covering all floor surfaces, and heavily upholstered seats. Comparing this hall to Carnegie Hall is useful in illustrating the point. Intuitively, the air volume of Radio City Music Hall is twice that of a concert hall, meaning that the fixed maximum sound energy of a symphonic or operatic ensemble is dissipated in the large space and leads to results of low loudness and weak sound impact. The walls and the ceiling surfaces are at vast distances from the listener, shaped in such a way that the sound is directed to the back of the hall. The rear walls of Radio City Music Hall kill any reverberation with their silk-screened fabric over thick fiberglass absorption materials.

The hall has excellent acoustics for amplified programs but has poor acoustics for unamplified performances. The reason lies in the specific and quantifiable acoustic qualities required for classical events. The acoustic qualities must be appropriate for the performance type. This will be proven and discussed in greater detail later, but for now, the point to be made is that excellent acoustics are defined only when the type of performance is fully defined (see Figure 2.2).

Reverberation Time

Opera houses and ballet halls share many of the same acoustic criteria for excellent sound but differ from concert halls or amplified music halls. The Seattle Opera House in Seattle, Washington, has a lower air volume than both Radio City Music Hall and Carnegie Hall and a lower ceiling. Excellent acoustics for opera and ballet are defined by different criteria from that of amplified music halls or concert halls. The acoustics of the Seattle Opera House would be considered only fair for symphonic performances. Compared to Carnegie Hall or Boston Symphony Hall, it has a lower reverberation time (RT) and shapes that optimize the balance of stage singing to orchestra pit musical instruments.

Consider the following optimal mid-frequency RT ranges for excellent acoustics in different spaces:

- Classical music (unamplified): 1.7–2.0 seconds
- Opera/ballet music (unamplified): 1.3–1.7 seconds
- Amplified music or speech: 1.0–1.4 seconds

All performance types will fall into these three main categories. Other RT criteria exist for liturgical music and film theaters, but these facilities are not classified as multi-use halls.

Sabine and Eyring

The acoustician regularly uses two RT formulas in the design of multi-use halls: the Sabine equation and the Eyring equation.

The Sabine equation relates RT, the time required for the level of reverberant sound to decay by 60 decibels (dB), to the volume (V) and total sound absorption (A), assuming that the shape of the room allows reverberant sound to be evenly distributed throughout. The even distribution of sound is critical for RT to be defined. This is why large stadiums, outdoor theaters, and small rooms will not have a defined RT per Sabine's formula.

Eyring is most useful when room shape or absorption materials will not allow even distribution of reverberant sound. Cinemas, drama theaters, and media studios have wall and ceiling sound absorption treatments covering much of the walls and ceiling therefore canceling even sound distribution. The RT in a multi-use hall that is set for amplified events will more closely follow Eyring than Sabine.

The total absorption in units of sabins is the sum of all surface areas of the room, each multiplied by its sound absorption coefficient. The coefficient is typically a value between zero and one. A value of zero describes an ideal reflector in which all energy is reflected. Materials that are nearly ideal reflectors include heavy plaster, concrete, and thick wood. A value of 1 describes an ideal absorber in which no energy is reflected from the surface. Materials that are nearly ideal absorbers include open windows and thick glass fiber panels. Absorption coefficient values are affected by materials and frequency of sound, so total absorption and RT are frequency dependent.

RT at low frequencies under 125 Hz is almost always longer than high frequencies because materials that absorb low frequencies are much less commonly used in construction. Carpet and curtains are mistakenly thought to effectively reduce RT but do so only at high frequencies. Low-frequency absorption is critical to creating excellent flexible acoustics in multi-use halls.

Explaining Sabine's Equation

Sabine's equation is RT = kV/A, where k is a constant value that depends on the unit of length used to express V (volume) and A (total sound absorption). This equals 0.049 seconds/ft. or 0.16 seconds/m. The RT is directly proportional to the volume of the room and inversely proportional to its total absorption. As a room's volume and size increase, the RT increases. For a room of constant volume, where absorption is added, the RT decreases.

The equation is based on empirical data gathered by Wallace Clement Sabine. Sabine measured the duration of residual sound after an organ pipe was played in the Sanders Theatre at Harvard University, as a function of the total length of seat cushions on the unoccupied room's benches. The data and the resulting equation were published in the January 1915 issue of the *Journal of the Franklin Institute* (Sabine 1915).

An important derivation of the RT equation is the mean free path (m). Sabine determined an approximate mean free path between successive reflections in a compact room as $m = 4V/S$, where S is the total area of the room. In 1947, A. E. Bate confirmed the accuracy of Sabine's approximation for a variety of compact shapes, as reported in the 1947 *Proceedings of the Physical Society* (Bate and Pillow 1947). This means that the mean free path will equal two-thirds of an edge for a cubic room, four-thirds the radius for a spherical space, and vary from two-thirds to twice the smallest dimension for a rectangular room.

Mean free path is often misunderstood even by seasoned acousticians. For example, rooms that are oddly shaped, very long and narrow, or very tall in relationship to its footprint will not follow Sabine's formula. Neither will a theater or concert hall with multiple small spaces and volumes off the main volume. Neither will a theater with volume broken up by a series of walls or partitions that reduce the distance that sound travels to at least two-thirds of the distance of the short dimension. This is often why halls with adequate volume fail to achieve the calculated RT.

Deriving Sabine's Equation

For those who are interested in why we use 0.049 times the volume in the Sabine RT formula, additional information is included below. Others who may be less mathematically inclined can skip ahead and trust that 0.049 is the correct number for English units.

The intermediate quantities involved in our derivation will be an average absorption coefficient, a, for the room, an average decay d per reflection (in dB/reflection), an average reflection rate R (in reflections/second), and an average decay rate r for sound (in dB/second).

The average absorption coefficient is simply

$$a = A/S.$$

At each reflection, on average, the fraction of energy lost is a, and so the fraction of energy reflected is $(1 - a)$. The ratio of incident sound intensity to reflected sound intensity is then $1/(1 - a)$, and the average change in sound level is

$$d = 10 \log (1/(1 - a)) \text{ dB/reflection.}$$

The difference in level between two sounds is defined to be 10 times the base 10 logarithm of the ratio of the sounds' intensities.

The average reflection rate is the distance traveled in 1 second (the velocity of sound) divided by the average distance between reflections (the mean free path):

$$R = c/m \text{ reflections/second.}$$

The average decay rate is the average reflection rate multiplied by the average decay per reflection:

$$r = \mathrm{Rd} = (c/m) \ 10 \log (1/(1 - a)) = (cS/4V) \ 10 \log (1/(1 - a)) \text{ dB/second.}$$

Adopting foot units for length, we plug in a value of 1130 ft./second for c and note the relationship that, for small a, $-2.3 \log (1 - a) \approx a$:

$$r = 1230 \ (S/V) \ (-2.3 \log (1 - a)) = 1230 \ Sa/V = 1230 \ A/V \text{ dB/second.}$$

Finally, if we want to define RT as corresponding to a 60-dB decay, the time required will be

$$T = 60/r = 0.049 \ V/A \text{ second.}$$

For metric units, the constant is 0.16.

The Eyring equation is as follows, where alpha, S, and V are identical to Sabine:

$$T_{60} = \frac{0.16V}{-S \ln(1 - \bar{\alpha})}$$

with

$$\bar{\alpha} = \frac{1}{S} \sum_i S_i \alpha_i.$$

I will spare the reader the derivation of this formula. Please refer to the glossary for more information about the technical terms and formulas often used in acoustics.

Keep in mind that a correct RT does not alone define excellent acoustics; it is but one very important factor in the design of halls. RT must be modified and adjusted in multi-use halls by adding sound-absorbing materials such as drapes, fabrics, or panels of absorptive materials on tracks or guides. Getting the RT on target is only one criterion that must be optimized for excellent acoustics. Halls with correct RT can suffer from echoes, dead spots, poor clarity, and lack of intimacy. Conversely, there are halls where the RT is on the modest side, yet the acoustics are outstanding. Carnegie Hall, for example, is at 1.8 seconds mid frequency compared to Boston Symphony Hall at 2.0 seconds.

There is much more to excellent acoustics than RT alone.

Chapter 3

Requirements for Excellent Acoustics

Introduction

Excellent acoustics come from the correct balance of many factors that are both interrelated and interdependent. Change one factor, and many others are affected positively or negatively. One architect I worked with explained, "It's a domino effect—change one small thing, and there are ripples that affect everything else." Technical terms stated here are further clarified and defined in the glossary located at the end of this book.

Reverberation Time

Reverberation time (RT) is the most important component of the design of a multi-use hall. If the RT is not correct, especially if it is too low for symphonic concerts, then the hall will be acoustically substandard.

Distribution of Reverberant Sound

Recall from Chapter 2 that Sabine's equation relates RT to the volume V and total sound absorption A of a room. However, reverberant sound is not evenly distributed in most halls that have balconies and upper level seating. Numerous measurements in halls clearly document higher RT levels in upper reaches of the room than on the main floor. Yet, ironically tickets and seats on lower levels are almost always priced higher.

Chris Jaffe, would claim that his goal was to make all seats in the hall sound like the balcony seats. In fact, he would jokingly go so far as to suggest elimination of the main floor altogether in favor of building only balcony seats.

RT is also perceived as lower on the main floor because the energy level of the direct sound is different. Close to the stage, a high level of direct energy emanates from the performers (loudness defined as G), and the relative level of RT is perceived as lower. Further away from the performers, RT is more in balance with the loudness level G.

Reverberation Level and RT

The energy level of RT is almost as important as its value in seconds. This is often misunderstood. If RT cannot easily be perceived, relative to background noise then what value does it have? A strong and vibrant RT is perceived as enveloping and richly surrounds the listener. Large concert halls with more than 2500 seats often exhibit this flaw. The RT is calculated to be 2.0 seconds using Sabine's equation, but the RT level is too low since the volume is large. It rapidly becomes inaudible, so a 2.0-second RT is perceived to be much less (see Figures 3.1 and 3.2).

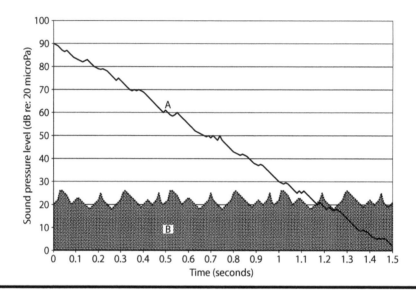

Figure 3.1 A reverberation time graph showing sound energy decay audible above noise floor. (A) Decay. (B) Noise.

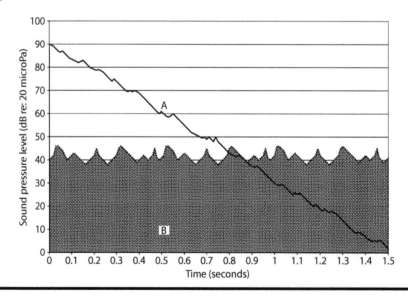

Figure 3.2 A reverberation time graph of the same hall showing the end of sound energy decay not audible above noise floor. This is perceived as less reverberant. (A) Decay. (B) Noise.

Running Liveness

Running liveness is defined as the reverberation that occurs within the music itself rather than the audible decay when the music suddenly stops. This factor is also known as the fullness of tone. Running liveness is composed of two parts including RT and early sound.

The RT in this case places an emphasis on the early part. The running liveness is the early sound that arrives from the musicians, which is defined as sound that arrives at the listener within 80 milliseconds of the direct sound. The ratio between the RT and the early sound is called running liveness or alternatively called fullness of tone. If there is too much RT sound energy, the sound will be indistinct and muddy, whereas too much early sound will lack fullness and life.

Reflections

Other acoustic criteria for a designer to consider include the direction, loudness, and frequency components of early reflections, midtime reflections, and later reflections, as well as diffusion and blending of the sound. Reflections are critical to increasing the intimacy and connections between performers and audiences. Each type of reflection is discussed in the following.

Early Reflections

Early reflections are the sound reflections that occur from surfaces near the sound source and arrive at the listener's ears soon after the direct sound. They can occur from walls, acoustic reflectors suspended overhead, or the stage floor and low overhead ceilings (see Figure 3.3).

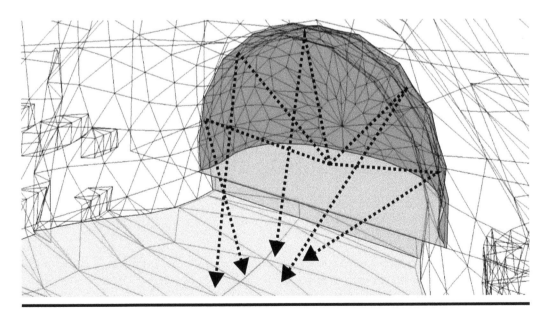

Figure 3.3   Globe-News Center, Amarillo, TX, 2006. A CATT digital model used for study of early (0–30 ms), mid (30–80 ms), and late reflections (80 ms and higher) from hall surfaces. Distinct reflections 90 ms after the original sound are destructive reflections.

For specular (image-preserving) reflections from flat surfaces, delays of more than 80 milliseconds can cause perception of a distinct echo. This delay corresponds to sound path length differences greater than about 80 ft. (24.4 m). Reflections arriving after the direct sound between 15 and 30 milliseconds are combined with direct sound to constitute early sound. These enhance clarity and definition and occur within the window of the initial time delay gap (ITDG). The definition of ITDG is further defined in the glossary.

The directions of these reflections are also important to acoustics. Research has shown that early reflections are best for acoustic quality when they arrive from the lateral plane in the cone of acceptance of about 15°–20° above horizontal. Early reflections from overhead or behind the listener are less desirable. Reflections of sound after 80 milliseconds from rear walls or ceilings can cause distinct echoes, sound odd to audiences, and become destructive to performers' sense of rhythm.

Clarity or C80

Clarity is subjectively defined as the ability for the listener to hear subtle and small details in music, such as the plucking of a string, known as pizzicato, or sung constants such as *P*, *T*, and *V*. It is widely accepted that the best way to determine if a space has clarity is to compare the logarithmic ratio of the sound arriving in the first 80 milliseconds after the direct sound with the sound arriving after 80 milliseconds.

C80 is not an independent variable and is highly correlated to both RT and whether a hall is occupied or not. Generally for multi-use halls, we target a C80 of zero to –2 dB occupied. Location of the audience in a hall will also determine C80 values with seats closest to the stage having higher values and naturally seats in the balconies' lower levels. Negative values mean that reverberant energy is greater than the early energy.

Early reflections combine with the direct sound to add strength to the sound source, whereas sound arriving after 80 milliseconds adds to the sense of reverberance. When a space has clarity, individual instruments can be heard from within a large orchestra, and speech is easy to understand.

When tuning a hall, I listen for the conductor to direct musicians while speaking in the opposite direction from the audience. If he/she can easily be heard in the audience, this indicates that the hall has a high degree of clarity.

Since clarity is enhanced by having early reflections within 80 milliseconds of the direct sound, the surfaces must be in the right location and angled in the correct orientation in multi-use halls. How to design and locate those surfaces is discussed in detail later, but the following guidelines highlight critical points.

Guidelines for Clarity

Early reflections for clarity must occur from surfaces that are specular, not diffusive. A slight bow to a surface and to surface treatments is acceptable, but clarity is enhanced when the reflecting surface is not highly diffusive nor articulated.

Clarity reflections need not be in full frequency in order to be effective. Only mid- and high-frequency reflections are required to give the impression of clarity. Low-frequency reflections less

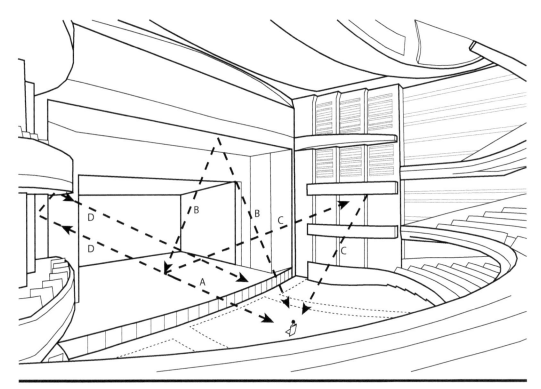

Figure 3.4 Dell Hall at Long CPA, Austin, TX, 2008. Clarity (C80) reflections at Dell Hall. (A) Direct sound from the musician to the listener. (B) Ceiling reflection at 15–20 ms after direct aides clarity. (C) Preferred sidewall reflections 40–80 ms after direct aids clarity and spaciousness. (D) Soffit reflections driving early reflections down into the audience.

than about 200 Hz add little to the impression of clarity. This means that lighter-weight reflectors made of thin materials such as glass, plywood, honeycomb, resin, or composites can be effectively used to achieve clarity in a hall. Using lighter-weight reflectors can represent huge cost savings in multi-use hall designs.

Side reflections also enhance spaciousness and source width and provide a sense of envelopment. When reflections come from behind the listener, they tend not to enhance clarity but do help with the sense of envelopment. This is called the surround sound effect (see Figure 3.4).

Spaciousness

Spaciousness refers to the broadening of the musical image and the dimensionality of the sound. Sound that lacks spaciousness seems to come from a single point on stage and does not sound full. It is only one or two dimensional in quality when a three-dimensional sound is desired. Spaciousness is directly related to clarity reflections but closely tied to reflections from lateral reflections off side walls. These reflections require side boxes, side balconies, or other surfaces that form a 90° angle between the wall and the soffit. This can occur from ceilings intersecting with the walls of a hall's upper level. This is explained in detail in Chapter 11.

Envelopment

Envelopment is the sense of being surrounded and immersed in the sound of a musical performance and has been identified in numerous studies as being critical to concert hall acoustics. Envelopment is feeling as though you are submerged in sound.

Guidelines for Envelopment

A large volume in a hall is critical to achieving envelopment. A large volume directly over the seating area such as in the attic space or within the roof truss is sometimes referred to as the reverberant cap or resonating volume. To be enveloped, the sound must swirl all around the listener as late reflections arriving from multiple directions, and there must be an adequate mean free path (see Glossary for definition). The RT must also be above 1.7 seconds.

Studies have shown that sound arriving at a person's left and right ears that are slightly different in time and from multiple directions adds to envelopment. This requires that the hall has diffusion surfaces that direct and mix the sound especially later in the reverberation period after 80 milliseconds after the direct sound. This can happen naturally of course unless you sit in the seat located in the exact center of the room.

Bass Ratio

Bass ratio (BR) is the sum of the RT of low frequencies (125 and 250 Hz) to the sum of the mid-frequencies' decay time (500 Hz and 1 kHz). A hall with well-respected symphonic acoustics has a BR between 1.1 and 1.4.

Halls with BR in the preferred range tend to impart a rich, warm sound to basses, cellos, low brass instruments, and percussion. For amplified music, high BR halls tend to be described as boomy for bass sound.

Guidelines for BR

To design halls with a good BR, large surfaces of walls must be solid and massive and typically built of brick, concrete, or other nondiaphragmatic absorbing material. Ceilings should have multilayer drywall or thick plaster surfaces and should not have large areas of thin wood or metal paneling that absorb low-frequency sound. In halls with less than 800 seats, this criteria can be relaxed and should have walls and ceilings built upon layers of drywall or plaster in order to achieve the massive surface weights required to achieve minimal low-frequency absorption.

In general, adjustable acoustic systems are not effective at reducing BRs for amplified music. This is due to the fact that the absorption coefficients of these materials are typically low between 125 and 250 Hz. This is covered in detail in Chapter 15 where various adjustable systems are described.

Diffusion

A hall with good diffusion sounds free of acoustic glare and harshness. There is no direct way to objectively measure and discern the amount of diffusion in a hall. Interaural cross-correlation, as defined in the glossary, and a comparison of RT in various parts of the hall can both be used to determine if there is an even dispersion of sound energy.

Guidelines for Diffusion

The determination of appropriate amounts, locations, and construction of diffusion surfaces constitutes a learning process. There are no silver bullets in terms of the perfect amount of diffusion.

However, it is clear that large areas of flat and planar surfaces on walls and ceilings are not conducive to good diffusion. On the other extreme, covering every surface with bumps and extensive surface treatments may actually serve to lower RT at high frequencies because too much diffusion becomes absorption. Ultimately, diffusion needs to spread around the multi-use hall, not just concentrate on one wall or on the ceiling.

It is best to avoid excessive diffusion on surfaces that are used for early specular reflections to increase clarity, C80. Small-scale diffusion does little for mid- and low-frequency diffusion, both of which are needed in order to achieve excellent sound.

I liken this to a recipe for soup. Many ingredients are needed, but high-frequency diffusion or the texture of a surface takes on the role of salt in soup. It makes it taste better, but other ingredients including mid- and low-frequency diffusion are also necessary in the recipe. Without salt, the soup will be less satisfying but will still be soup. Diffusion of all frequencies makes for a more satisfying listening experience (see Figures 3.5 and 3.6).

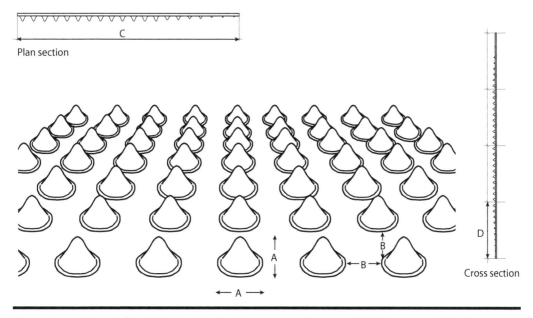

Figure 3.5   Alice Tully Hall, New York, NY, renovated 2009. High-frequency diffusion cones. (A) Height of cone and base diameter varies. (B) Constant spacing between cones. The diffusion covers an area 16 ft. (4.8 m) wide by 12 ft. (3.6 m) tall on upstage wall of stage area and on hall sidewalls. (C) Plan section. (D) Cross section.

Figure 3.6 Alice Tully Hall, New York, NY, renovated 2009. Molded glass reinforced gypsum high-frequency diffusion cones on upper sidewalls. (Photographed at Lincoln Center for the Performing Arts in New York City.)

Loudness

This factor is defined as the strength of the arriving sound energy or the intensity and impact of sound on the listener. The loudness or acoustic gain in a hall is easily measured but is only moderately important in the design of multi-use halls with a reasonable seat count.

Theory indicates that as RT increases, so does G, as there is more energy from reflections, and the strength is increased with the support of the reverberation. Halls with very large volumes and seating capacity tend to have lower G values. Beranek (1996) states in *Concert Halls and Opera Houses: How They Sound* that most concert halls have G values equal to 2–5 dB. Halls with G near 5 tend to be overpowering, and those with less than 3 are weak sounding.

Guidelines for Loudness

Multi-use halls with more than 2400 seats will have low G as the volume is too large for the sound of unamplified voices or instruments. Wide, fan-shaped halls or surround halls should not exceed 2000 seats, or G will suffer. Multi-use halls with fewer than 500 seats, low volume, and poor stage design can have too much G, resulting in a harsh sound that is overly bright.

If the stage end of the room is designed as a megaphone, G levels will be high, but balances between instruments will suffer. This in turn reduces acoustic quality. Sacrificing G for a gain in balance can be worthwhile—remember that some of the best seats in concert halls are in the balcony where G levels are low.

Stage Sound

The acoustics of a hall are only as good as the quality of sound that leaves the performance area. The acoustics in a multi-use hall have no hope of being excellent unless the sound coming from the stage is well balanced and blended with proper loudness.

Guidelines for Stage Sound

Musicians must be able to hear themselves and other sections of the orchestra clearly in order for them to gage their playing and perform as an ensemble. The sound coming from the stage must be well blended so that one instrument or section is not overemphasized or unbalanced. Soloists must be heard over the ensemble and not overwhelmed by loud instruments.

The stage or pit should never be considered a megaphone or loudspeaker that only sends sound out into the hall. The stage is an integral part of the overall sound in the room. Remember, sound moves quickly and covers a lot of ground in 2 seconds—about one-third of a mile or 2200 ft. (600 m).

An orchestra shell is a critical part of a multi-use hall's success in regard to classical and unamplified music. The shell's design, shape, features, and flexibility are the key to a successful multi-use hall (see Figure 3.7). We devote an entire chapter, Chapter 10, to this complex topic.

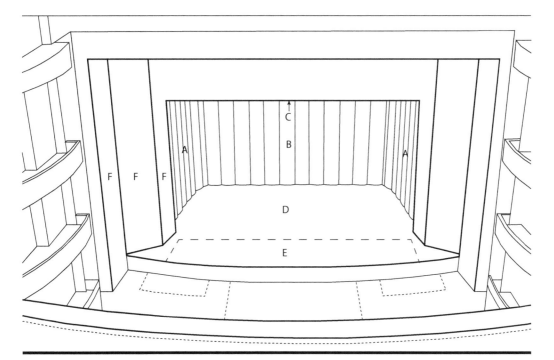

Figure 3.7   Dell Hall at Long CPA, Austin, TX, 2008. (A) Rolling sidewall shell towers have diffusive convex shape and taper gently upstage. (B) Rolling rear wall towers, similar to sidewalls, roll off stage for storage. (C) Shell ceiling (not shown) discussed in Chapter 9. (D) Orchestra sound source area includes both musicians and chorus. (E) Forestage apron area is part of the orchestra shell design for string section and soloists. (F) Forestage walls and ceilings must work in concert with the shell design.

Noise and Vibration

Contrary to common belief, the acoustician's concern about reducing vibration and noise in multi-use halls is not about reducing annoying or disturbing sound. Our concern is that background sound destroys the quality of the experience. Noise masks subtle acoustic qualities and renders them inaudible. Poor halls are often good halls with excessive noise.

Guidelines for Noise and Vibration

Multi-use halls need to be very quiet in order to achieve acoustic excellence. There is a point of diminishing return (typically NC-15) beyond which dropping noise levels toward the threshold of hearing, often called NC-1, is unnecessary. The definition of NC can be found in the glossary.

It is almost impossible and extremely costly to fully isolate all outside sound. The acoustician should aim to achieve a reasonable amount of sound isolation with a small percentage of outside sound heard, say, 5%–10%.

Sound leaking from one performance space and into another is never acceptable. If performance spaces are to be used simultaneously, total sound isolation between spaces must be achieved.

Structural breaks and acoustic joints are often used to achieve sound isolation between loud performance spaces and noisy mechanical rooms. These are used because sound and vibration travel relatively easily through common structures under certain conditions. There are a number of excellent resources on the design of very quiet air systems and sound isolation constructions, so this will not be discussed in this book. You may want to reference the following texts:

- *Acoustical Designing in Architecture* by Vern O. Knudsen and Cyril M. Harris (1950, 1980)
- *Noise Reduction* by Leo Beranek (1991)
- *Acoustic Design and Noise Control*, Volume II by Michael Rettinger (1977)

Chapter 4

Creating the Acoustic Program

Introduction

An exciting commission for a new multi-use hall has been awarded to the design team, which includes architects, acousticians, theater consultants, and engineers. What happens first?

Buildings are successfully designed not by an individual but by the shared vision and expertise of a team of experts working in collaboration and harmony with an owner, user groups, donors, and stakeholders. The best multi-use halls come from a process where the entire team is engaged right from the beginning of the project, not brought in after the design is set. The programming phase is the process where the uses of the hall are identified, priorities are set, and the audience size is discussed. Building costs and budget are also discussed in this phase.

The programming of a multi-use hall is an exhilarating and enjoyable task. The creative imagination is fully engaged in union with the team, and the process really draws on the experience of the acoustician. The goal is to come up with the right mix of spaces to meet the needs of all stakeholders and stay within the target budget. The latter is critical.

Programming Phase

During the programming phase, the acoustician listens, asks questions, and makes suggestions but does not dictate a solution or design. Listening to what is wanted and asking probing questions are essential during programming. Active listening involves asking open-ended questions and then creating a plan for the space that meets those needs and can be built within budget. The acoustician does not lead the programming effort but supports the effort by focusing on the areas of importance to the acoustic performance. Some of these areas include symphonic use, orchestra pit, orchestra shells, and adjustable acoustic needs.

Programming for Renovations

Thoughtful attention to the acoustics of an existing space that is to be improved and renovated begins with a series of walk-throughs and site measurements. Here, acousticians do not have the luxury of a blank canvas but instead inherit the challenges involved with an existing site, building, structure, systems, and adjacencies. Appreciative observation of the building's assets and liabilities is vital to assisting the design team and the stakeholders with programming development.

Tools

A questionnaire distributed to the potential user groups is a tool that is very useful in gathering information. An example of a programming questionnaire used for a multi-use hall and performing arts center at a college is included at the end of this chapter.

On-Site Measurements

Acoustic measurements of the hall should be taken as soon as possible to certify reverberation time (RT), bass ratio, C80, and other key attributes. Documenting the noise levels of existing heating, ventilation, and air-conditioning (HVAC) systems is important. If there are multiple systems in the hall, then each should be run both individually and collectively to determine the noise contributions.

I prefer to begin by measuring the noise level of a hall with all HVAC systems off. This makes it easy to determine if there are other noise sources that should be considered such as elevators, escalators, or outside mechanical gear. In one project in New York City, the noise level in the hall was negatively affected by a neighboring building's water pumps from the other side of the alley. Their pump was rigidly mounted to the floor with no isolators. The noise crossed the alley through common foundations and was clearly audible on stage. Diagnosing this noise was only possible after the process of shutting down every system in the building.

Site noise and vibration from air, vehicular, and rail traffic can be a major concern when programming a multi-use hall. Noise must be witnessed in person and measured through site noise surveys as well as by reviewing data from sources such as the Federal Aviation Administration, railroad companies, and subway systems.

Alice Tully Hall Example

Alice Tully Hall in New York City is situated only a few yards from subway trains that run up and down Broadway. Initial measurements determined that there was a 100–150-Hz rumble from passing trains. This was only audible when HVAC systems for the hall were shut down, and it was very obvious when the NC-25–30 noise from the high-velocity air system was cut off. Users of the hall were surprised when I shut off the air and let them fully experience the impact of the Number 2 train. They had played and listened in the hall for years and never realized that the noise had been there all along.

Reducing this noise was an important programming goal of that project. The need to develop solutions was threatened by cost cuts many times; however, the team prevailed in preserving noise mitigation systems. For more on this story, read the Alice Tully Hall case study (Appendix I).

There are many excellent texts and resources on how to measure site noise and vibration, and they will not be covered here.

Look for Opportunities

When surveying an existing hall for renovation, pay particular attention to the places that are often dismissed as unimportant, such as the attic or area above the ceiling and below the roof. The attic is an area where additional acoustic volume can be added to the hall to increase RT. It also can be the ideal location for adjustable acoustic banners, drapes, or panel systems (see Figures 4.1 through 4.3).

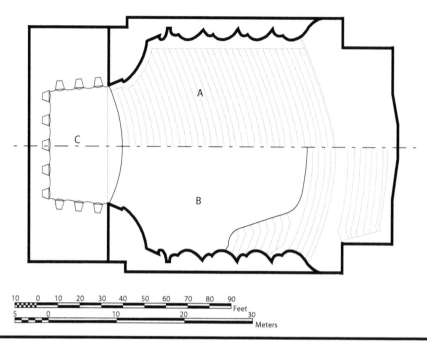

Figure 4.1 **Bass Concert Hall, University of Texas, Austin, TX, renovated 2007, 3000 seats.**
(A) Orchestra level lacked early reflections for wall surfaces. (B) Balcony level covered large areas of
the orchestra seating. (C) Symphony played in orchestra shell as apron had no supporting surfaces.

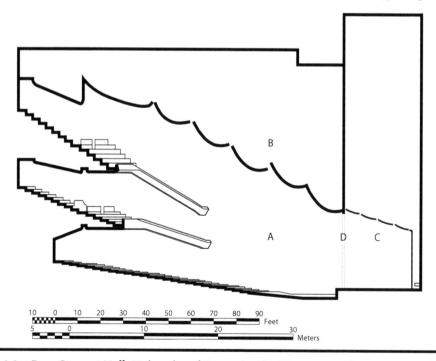

Figure 4.2 **Bass Concert Hall, University of Texas, Austin, TX, renovated 2007. (A) Small acous-**
tic volume for 3000 seats, limited RT. (B) Large unused attic space closed off from hall by plaster
ceiling. (C) All musicians played in orchestra shell behind proscenium arch. (D) Proscenium arch.

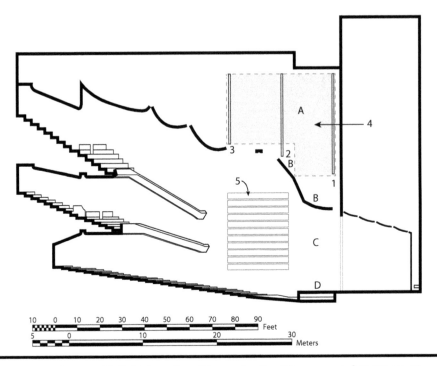

Figure 4.3 **Bass Concert Hall, University of Texas, Austin, TX, renovated 2007. (A) New acoustic volume exposed by removing old ceiling and adding new forestage reflector (B), with sound transparent ceiling. (C) Side walls reshaped for early lateral reflections and diffusion. (D) Stage lift now used for orchestra string section and soloists. One to five new adjustable acoustic drapes in attic and on side walls.**

Example: Secrets in the Attic

At Bass Concert Hall at the University of Texas (UT) in Austin, Boora Architects, Auerbach Pollock Friedlander theater consultants, and Jaffe Holden were called in to program the renovation of the lobby and the hall. The initial walk-through identified a huge attic space above a heavy plaster hall ceiling and a very tall roof. During programming sessions, the music department remarked that the hall was too dry and harsh sounding, indicating a too-low volume. The management complained that the hall was too live for amplified Broadway shows and amplified music, a mainstay of their programming. A new program direction was prepared for the hall after a review of management's input, original construction drawings, and our acoustic modeling efforts.

The ceiling was surgically removed in the front of the hall where it would have the greatest impact. This was actually an inexpensive upgrade. By removing the front portion of the ceilings in the hall and adding in a new smaller forestage reflector and rigging positions, usability for touring productions was enhanced, and adjustable acoustic systems could be added easily into the large attic volume.

Motorized velour curtains on simple traveler tracks could be extended or retracted. This provided a range of adjustability to the acoustics of the hall so that it could be more reverberant for symphonic productions and drier and controlled for amplified events. Manual pull drapes were also added to the reshaped lower cheek symphony reflection walls to reduce reflections from suspended line array speakers hung left and right of the proscenium when in use (see Figures 4.3 and 4.4).

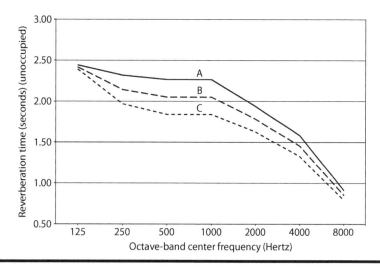

Figure 4.4 Bass Concert Hall, University of Texas, Austin, TX, renovated 2007. Reclaiming acoustic volume above the ceiling raises RT for the symphony. Adjustable acoustic drapes above the ceiling can lower RT for amplified productions below the existing hall's former RT, enhancing amplified programs. (A) Post-renovation drapes/banners stored. (B) Pre-renovation. (C) Post-renovation drapes/banners deployed.

Reduction of the NC-20–25 HVAC system noise was not in the work scope, and changes to the overdiffusive wall shaping were not feasible. However, these modest acoustic changes did make a substantial difference at the Bass Concert Hall, and the symphonic acoustics much improved.

Advocate for Musicians

Most projects have a modest budget, so an important element of the acoustician's role in programming is to advocate for the musicians when there is a desire to cut costs and area. Acousticians must strive to ensure that the specific needs are met for all musical groups ranging in size from soloists to a large ensemble (see Figures 4.5 through 4.7. This hall is also featured in a case study).

Programming the Stage

The stage and the orchestra pit are two areas in programming where the acoustician's input is vital. Getting the size right (shape, width, depth, and height) is a critical first step, as most musicians rehearse in the space in which they perform. This strategy is discussed in more detail in Chapters 9 and 10.

Orchestra Stage Position

Orchestra stage position is very important, and its programming is driven by the needs of the symphonic orchestra. There are three stage positions: fully in the stage behind the proscenium, string sections forward of the proscenium on a lift, and orchestra on two or three lifts in front of the proscenium. The decision of stage position has a profound impact not only on the stage, lift, and storage requirements but also on the hall itself. Its position must be solidified early on in the process. Chapter 10 features a comprehensive discussion of the space and programming needs of the stage, and an overview is included in the following.

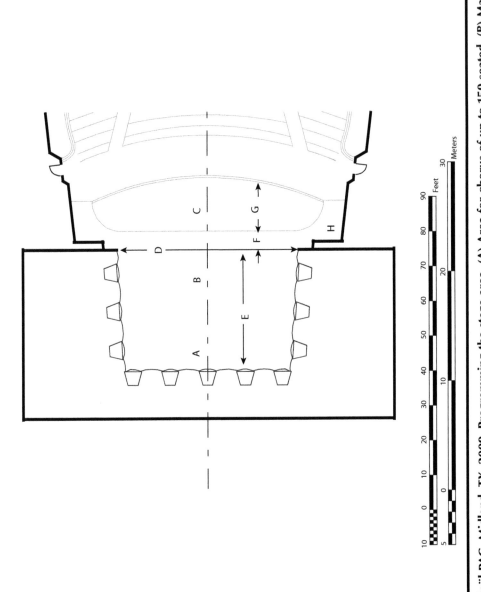

Figure 4.5 Wagner Noël PAC, Midland, TX, 2009. Programming the stage area. (A) Area for chorus of up to 150 seated. (B) Main area for orchestra of up to 100 musicians that will extend into forestage area C. (D) Proscenium width 52–54 ft. (15.8–16.4 m) to 60 ft. high (18.2 m) depending on shell configuration. (E) Shell depth of 40 ft. (12.1 m) minimum. (F) Apron 4–6 ft. (1.2–1.8 m) and a pit lift/stage extension of 9 ft. (2.7 m) to 26 ft. (8 m), depending on shell configuration (Chapter 10). (H) Throat walls. See Chapter 12 for details on this area.

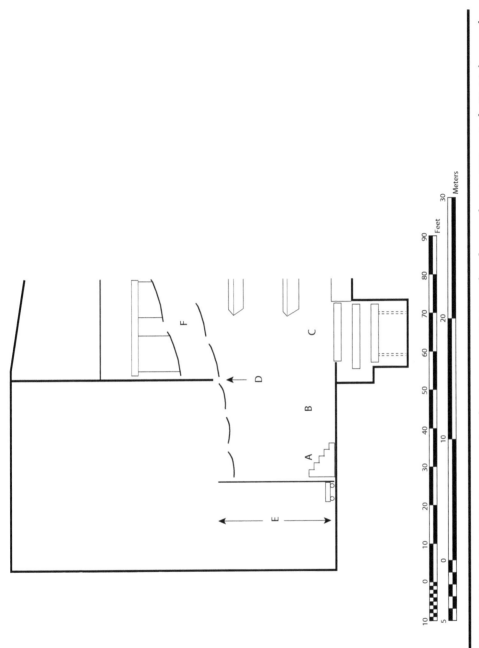

Figure 4.6 Wagner Noël PAC, Midland, TX, 2009. Programming the stage area. (A) Area for chorus of up to 150 seated. (B) Main area for orchestra of up to 100 musicians that will extend into forestage area C. (D) Proscenium height 30–35 ft. (9.1–10.6 m). (E) Shell towers typically 30 ft. (9.1 m) tall. (F) Forestage reflectors, fixed or movable from 30 to 40 ft. above the stage (9.1–12.2 m).

Figure 4.7 Wagner Noël PAC, Midland, TX, 2009. Large shell programmed to serve the local symphony orchestra with a large chorus. Forestage ceiling can be raised and lowered. Featured as case study.

Proscenium

A width of 60 ft. (18.2 m) is suitable for symphonic orchestras up to 100 members when positioned behind the proscenium. If the orchestra will play 15–20 ft. (4.5–6.1 m) forward of the proscenium line on a pit lift, then this dimension can be reduced to a minimum of 52–54 ft. (15.8–16.4 m). The proscenium wall and opening can act as a throttle, or choke, that restricts the free flow of sound from the orchestra to the hall and back to the musicians.

Stage Performance Area

A good rule of thumb is to allow 24 ft. (2.2 m²) per professional musician for the symphonic stage area. College musicians can be in a tighter area (–10%), and high school students can be in an even smaller space (–20%), but 2400 ft.² (222.9 m²) for a symphonic orchestra should be the starting point. A depth of 40 ft. (12.1 m) from stage edge to upstage wall by 60 ft. wide (18.2 m) generally works. The stage area needs to accommodate the concert enclosure or shell with these minimum dimensions plus the tower base dimensions. The apron should be 4–6 ft. (1.2–1.8 m) deep and can then serve as valuable added playing area for the stage (see Figures 4.5 and 4.6 for example).

Chorus Space

Communities with extensive choral programs require a stage and shell that can house the full orchestra in addition to the chorus of 100 to 200. Since the chorus is usually seated for many sections of a symphonic piece, there must be adequate space for risers. Allocate about 6–8 ft.^{2} (0.8–0.9 m) per seated chorus member, spaced about 2 ft. (0.6 m) apart. This may lead to an exceptional large stage and orchestra shell; however, the overall depth of the stage is the critical component of multi-use hall design (see Figures 4.5 and 4.6).

Back of House Space

Do not forget to petition and plan for adequate storage space in the back of the house. Shells and pianos require storage, musicians often require changing rooms, and a rehearsal room is often required.

Shell Storage Space

An adequate shell storage area ranges from 200 to 400 ft.^{2} (60.9–121.9 m^2) and has a minimum height of about 30 ft. (9.1 m). The necessary space really depends on how the shell is designed. This process is detailed in Chapter 10.

Locker Rooms

Changing rooms and locker rooms vary in size. In some communities, musicians arrive dressed in concert attire, whereas in others, they come in street clothes and change, dress, or even shower on site. A hall that houses a symphonic ensemble will have greater demands than a hall that serves as a road house that is rented to the ensemble for limited rehearsals and performances.

Rehearsal Room

A rehearsal room is the required space for use by the symphony, chorus, and performance groups. This should be a large room with a floor area of about 2400 ft.^{2} (731.5 m^2) so that it can function as a location for ensemble rehearsals and warm-up sessions. Often, these rooms have too low a ceiling. At an absolute minimum, use 25 ft. (7.6 m) to the underside of the acoustic volume. Keep in mind that no matter how skilled the acoustician, the rehearsal room is not acoustically sufficient as a substitute for the stage. The rehearsal room should be used for some, not all, rehearsals and is more suitable for sectional or choral rehearsals.

Instrument and Riser Storage

Piano storage in a humidity-controlled room directly off the stage is needed for one or two large grand pianos. Remember to also include storage space for large tympani, bass drums, and mallet instruments. Musician's stands, chairs, and risers also require storage. A full symphonic riser system requires major storage covered in detail in Chapter 10. Most multi-use halls need storage for a modest set of portable rectangular risers.

Orchestra Pit

Designing the orchestra pit is complex and depends on spaces located below and in front of the stage such as the trap room, rigging arbor pit, stair access, and storage areas. Pit design is covered in detail in Chapter 9. Programming the orchestra pit is much less complex.

Programming the Pit

The pit should be programmed so that 70%–80% of the musicians are out on the lift and not beneath the stage overhang. To increase width from left to right, extend it to the proscenium opening width. This will minimize the necessary depth. Plan for a modest 6 ft. (1.8 m) minimum deep zone under the overhang for the full width of the pit to the left and right. This allows for circulation and overflow for large ensembles.

Allow 17 ft.^{2} (1.5 m^2) per musician minimum for pits that hold 30–60 musicians. Increase the size to a minimum of 20 ft.^{2} (1.8 m^2) per musician for smaller pits of less than 30 musicians. These are rules of thumb for programming and should always be confirmed later in the design process.

Large opera companies require the most space: up to 110 musicians in the pit with 17 ft.^{2} (1.5 m^2) per musician. This might mean two or three stage lifts in addition to more extensive space under the stage. Regional opera and ballet companies tend to use a smaller pit. Plan for up to 70 musicians on a single lift.

A pit and lift area that is too large will waste client resources. One that is too small will affect the quality of productions. It is important to get the sizing right.

Balconies

Configuration of the hall's balconies, lifts, and side boxes will impact the program and concept of the building. Budget will also influence their design. For example, side boxes and side balconies are acoustically and visually constructive, but they are not particularly efficient seating areas and often require additional exits, circulation corridors, and stairs. This in turn impacts building cost. A higher number of balconies require taller stair towers and more elevator stops and lobby space (see Figure 4.8).

Programming the Balconies

Seating capacity and stage configuration drive balcony arrangements. For example, the 2300-seat Mead Theatre within the Benjamin and Marian Schuster Performing Arts Center in Dayton, Ohio, has three shallow wraparound galleries. These allow for excellent site lines and acoustic connections to the three large stage lifts upon which the Dayton Symphony plays. This is a very intimate and delightful room, but the inclusion of three galleries contributed to the construction cost.

The Long Center for the Performing Arts in Austin, Texas, has 2400 seats and two large balconies. These balconies are sound transparent and allow for more seats to be located under the balcony than might otherwise be acoustically acceptable. The Long Center has very modest side balconies with only a few seats as opposed to the Schuster Center's side boxes, which contain many more seats. This decision was primarily driven by cost. Budget was a primary factor in the decision to eliminate side balconies and boxes for the 1800-seat multi-use hall at the Southern Kentucky Performing Arts Center in Bowling Green, Kentucky.

Generally speaking, multi-use halls with fewer than 400 seats do not require a balcony. Halls with 400–1200 seats do well with a single balcony, and those with more than 1200 seats require two or three balconies.

HVAC and Sound Isolation

Site noise and vibration are a growing concern in the design of a multipurpose hall since quiet sites are rare, and urban sites are the norm. It is now vital to utilize sound isolation techniques

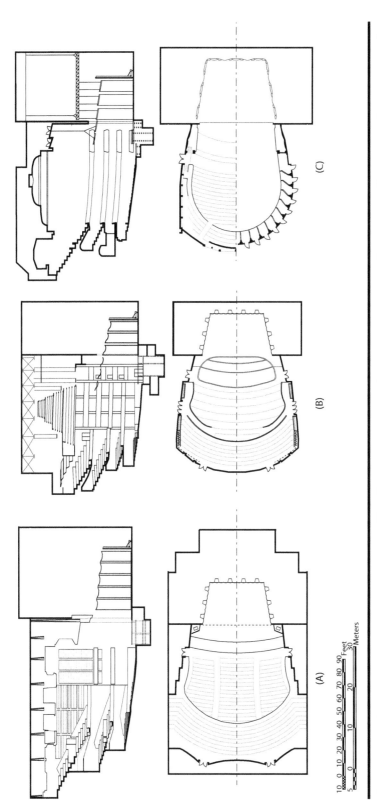

Figure 4.8 Schuster PAC, Dayton, OH, 2003; Dell Hall at Long CPA, Austin, TX, 2008; Bass Performance Hall, Fort Worth, TX, 1998. Plan and section comparisons for 2-, 3-, and 4-level multi-use halls. (A) Dell Hall, 2400 seats. (B) Schuster Center, 2300 seats. (C) Bass Hall, 2000 seats.

for exterior noise and vibration, and they should be included in budget and cost models. How to approach this challenge and how much isolation to provide must be addressed in programming and schematic design phases.

Site Noise and Vibration

There are a number of reputable texts that explain the measurement and documentation of environmental noise and vibration from transit systems. The reader is directed to texts such as *Acoustical Designing in Architecture* by Knudsen and Harris (1950, 1980) for complete directions on how to measure them. Documenting site noise and vibration conditions for all sites, especially urban locations, is an important step to take early in the process so the team has real data with which to work.

University Campuses

College campuses, except those downtown in urban areas, tend to need less robust isolation systems because they are in quieter locations far from transit, airports, and rail lines. Still, be aware of the hall's proximity to student unions or outdoor campus performance venues that could cause an impact. Low-flying aircraft and lawn maintenance machinery can also create disturbances that may require sound isolation.

Urban Sites

Urban sites are prone to general city noise such as traffic, emergency vehicles, and special events like parades or fireworks. Be aware of the location of hospitals, fire stations, and police stations. Determine if the hospital has a helipad, which will require extra sound isolation.

Aircraft

Identify nearby airports early on in the process. A hall that is directly on axis with a runway will have problems. Noise does drop quickly if the hall is off axis with a runway. Also, perform a check for military aircraft in the area (Figure 4.9).

Rail

Rail lines also cause noise and vibration issues. Noise emanates from train horns, locomotive rumble, steel wheels, and track noise and sends vibration and noise into the building structure. The solution for this issue requires significant structural isolation and box-in-box construction that is beyond the scope of this discussion. First, examine the possibility of isolating the track, as it may be more cost effective than isolating the entire building. Other buildings in the vicinity will most likely benefit from this isolation as well.

Subway lines and transit systems also present a complex vibration challenge. Refer to the Alice Tully Hall case study (Appendix I) for an example of how that problem can be approached and solved. Noise and vibration drop rapidly with distance from transit lines, and picking a more distant site could potentially save money (see Figure 4.10).

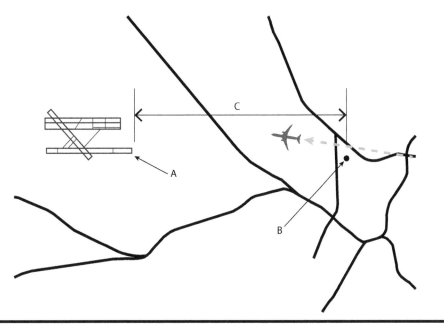

Figure 4.9 Dallas City Performance Hall, Dallas, TX, 2012. Programming and proximity to airport. (A) Dallas Love Field. (B) Dallas City Performance Hall. (C) Although 5 mi. (8 km) away, the hall is directly under the flight path. Featured as a case study.

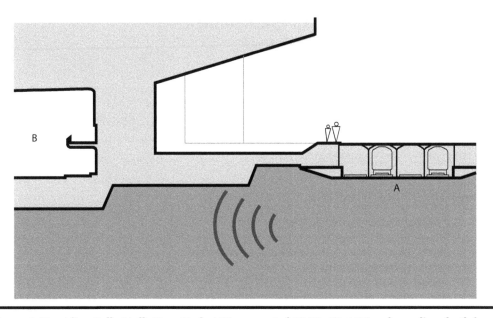

Figure 4.10 Alice Tully Hall, New York, NY, renovated 2009. (A) MTA subway line, both local and express tracks, are within 20 ft. (6 m) of (B) the hall. To reach NC-17 required high level of vibration isolation both to the hall and to the subway tracks. Featured as a case study.

Establishing Noise Isolation Criteria

The amount of sound isolation is defined mainly by budget criteria and expectations of owners and users. It is a rare performance space indeed that can afford total sound isolation. In rare instances, complete isolation might be necessary, but in most cases, isolation should meet the programmed needs and no more. The criteria must relate to audibility and the ability to feel vibration rather than a perceived ideal level.

Programming Background Noise Levels

A realistic isolation level for the building must be determined. Establish a reasonable background noise level (NC) that is both achievable and realistic for the type of program being presented and the expectations of the owners. Overdesigning and using unrealistic criteria only impede design, reduce the team's confidence, and add excessive costs. Discuss the expected number of noise intrusion events and the severity level of such events with stakeholders before embarking on the next phase of design.

Preparing the Guide

At the conclusion of the programming process, the acoustician prepares a programming or predesign phase summary. This is a report that discusses the decisions made and the rationale behind them. This document serves as a guide for the acoustic design as the process moves forward and must be approved by stakeholders and users.

Programming Requirement Questionnaire Example: County College District

A questionnaire distributed to potential user groups is very useful in gathering information. An example of a real programming questionnaire used for a multi-use hall and performing arts center at a college is included below. The Visual and Performing Arts Center is anticipated to include the following areas: a multi-use hall, lobby, back-of-house spaces, black box theater, dance studios, drama and art spaces, and gallery. Our firm has the following questions regarding the intended use of these spaces and their associated systems:

Acoustics/Theater—General

What type of programming will the space be used for?

_____

What types of programming should drive the design for the space?

_____

How often do you anticipate each of the following?	Never	Sometimes	Often
Band/orchestra performances	0	0	0
Quartet/quintet/solo instrumental performances	0	0	0
Choral performances	0	0	0

Lectures	0	0	0
Opera	0	0	0
Musical theater	0	0	0
Movie screenings (requiring surround sound)	0	0	0
Rock concerts	0	0	0
Theatrical productions	0	0	0
Video/data presentations	0	0	0
Film presentations	0	0	0
Touring artists/productions	0	0	0

Music Department—General

1. What will be the largest ensemble to occupy the theater on the stage or in the orchestra pit?

2. What is the size and composition of this ensemble?

3. What is the current annual number of performances?

4. Do your large ensembles currently rehearse or perform with floor risers?

5. What are all the main performance ensembles and their respective sizes? Also, what unique instrumentation is associated with any of the ensembles?

6. What are all the main choral performance ensembles expected to play here and their respective sizes?

Theater/Dance/Arts Department—General

What would be the largest production to occupy the theater?

What is the current annual number of performances by department?

How might these numbers change in the future?

_____

_____

What types of event productions would be in the space?

_____

_____

What joint programs do you have with the music department or other departments in the school?

_____

_____

Issues of Concern

1. Is there a current space that we should be aware of as a point of reference acoustically? Or are there other theaters that you find acoustically pleasing that we should know about?

_____

_____

2. What are your greatest concerns relating to acoustics, sound isolation, background, and audio systems in the theater?

_____

_____

CREATING
THE BUILDING

Chapter 5

Translating Program into Bricks and Mortar

Introduction

How do we take programming guidelines and translate them into a physical hall that engages the audience and performers? Acousticians do not need to reach for the crystal ball … there is a process to follow in designing the hall.

Creative Tension

I have spent more than 35 years thinking about the nature of the creative process in terms of creating a visually and sonically inspiring facility that functions at the highest possible level. I believe that when an acoustician, an architect, and a theater consultant work collaboratively together, remarkable results can be achieved regardless of budget.

It is also important that the owner or the stakeholder group can make tough decisions on program, budget, and schedule in a timely and forthright manner. This dynamic causes a creative tension where competing visual, acoustic, and theatrical priorities can be discussed in design meetings and over electronic media in an effort to produce a holistic hall design. A talented and committed team can resolve opposing interests to create solutions far superior than one team member could create alone. This is what makes the work fascinating to watch in action and even more terrific to be a part of.

The "I" Word

A key ingredient of great flexible halls is the creation of performer/audience intimacy. It is a connectedness that is palpable and visceral. The nature of a live group experience is fundamentally different from that of an audience at home. Making this experience special, inspiring, and memorable is the ultimate goal of the design team and the acoustician. To imagine new ways to create this outstanding experience, we must tap into the creative unconscious, utilize new technology, and investigate what other designers in the field are doing. The acoustician must listen to a variety of flexible and fixed-use performance spaces and glean the best aspects and experiences in order to understand how they were physically accomplished. The creation of an intimate sonic experience is accomplished by taking the broad range of other designs, combining that with one's own experience, and then collaborating with the design team and stakeholders.

Ideal Hall

A well-known theater architect once asked me to describe my ideal acoustic design for a hall. Frankly, I was unable to answer for a few moments. After some thought, I realized that there is no ideal acoustic design. A hall that has extraordinary acoustics but is visually uninspiring will rank lower than a gloriously beautiful space with equally impressive acoustics because audiences and performers listen with their eyes as well as their ears. Intimacy is both a visual and sonic experience—there is no intimacy in acoustics when great distances detach the performers or other audience members. I never have an ideal hall design in mind at the onset of a commission. Rather, intuitive factors based on experience and research are united with the architect's visualization and the theater consultant's contributions.

Educating the Team

There is great value in taking the team on a tour of other facilities that are similar in design, program, or budget so that a frame of reference and a common language can be developed. It is exceedingly useful to have this common language during the design process and especially when tough decisions must be made.

For example, the acoustician can reference the balconies at XYZ hall or the orchestra shell at XYZ hall so that the group can determine what they want in their hall. Experiencing the equipment in operation makes a far greater impact than discussing the device in a conference room. This speeds the decision-making process and provides a clearer understanding of complex issues. It also allows the touring committee to express their personal experiences in touring halls to the larger stakeholder group. This sharing of first-hand experience is much more successful in garnering buy-in than an arbitrary and unfamiliar bullet list from the design team.

Most importantly, trust is established and relationships are forged when a design team interacts with the owner's group during a tour. This is essential in moving forward in the design process.

Tips for Tours

- Call the technical director of the hall before you go to let them know that you would like to come and that you want a professional tour. Express that you want to go everywhere in the hall, from the pit to catwalks.
- Ask the facility to be ready to demonstrate acoustic devices for the tour such as the shell, acoustic drapes, and banners.
- Ask the facility to be ready to answer questions for the tour group.
- Have the manager or executive director of the hall be available to discuss operations, staffing, revenue, costs, and future plans.
- Charter a private plane or transportation so that more halls can be seen in a limited time frame by people with limited availability.

Acoustic Design in Preschematic Phase

There may not be a perfect acoustic design for a multi-use hall, but there are a number of critical parameters that must be established in order to achieve impressive results. Different performance types require quite different parameters involving reverberation, clarity, warmth, and bass ratio in order to impress audiences and performers.

It should be noted that the unamplified symphony's criteria are the most stringent, as they require the longest RT and a very quiet room completely isolated from internal and external disturbances. Unamplified Western opera, chamber music, and unamplified drama also drive design with demanding acoustic criteria, but no matter what, the acoustic design process begins by establishing symphonic requirements.

Symphonic Requirements

Symphonic requirements can be difficult to achieve and must be determined early on in the preschematic phase. Symphonic requirements affect all later decisions and can lead to costly changes if they are not addressed. The chart lists recommended acoustic volume (*V*) requirements in cubic feet per person as well as significant large multipurpose halls for comparison (see Figure 5.1).

Volume and RT

It is important to discuss volume per person (*V*/seat) early on in the programming process as this parameter is often misunderstood and misused. The *V*/seat value is used as a rule of thumb and not a strict recipe to establish broad outlines of the dimensions of a hall, ceiling height, and roof height. During the design process, the *V*/seat is adjusted and fine-tuned as just one parameter in the process of modeling a hall (see Figure 5.2).

Name	Location	Seat Capacity	Volume (ft.^{3})	Volume (m^3)	Volume/Seat (ft.^{3}/Seat)	Volume/Seat (m^3/Seat)	Mid-Frequency Reverberation Time (Seconds)
Alice Tully Hall	New York, NY	1100	371,180	10,511	337	9.6	1.6–2.1
Annenberg CPA	Los Angeles, CA	500	228,000	6456	456	12.9	1.2–1.7
Baldwin Hall at Duke	Durham, NC	685	271,000	7674	396	11.2	1.5–2.0
Bass Hall–Fort Worth	Fort Worth, TX	1900	720,000	20,388	379	10.7	1.7–2.0
Carnegie Hall	New York, NY	2804	883,000	25,004	315	8.9	1.8
Daegu Opera House	Daegu, South Korea	1500	388,000	10,987	259	7.3	1.2–1.6
Dallas City Performance Hall	Dallas, TX	750	362,000	10,251	483	13.7	1.9–3.3
Dell Hall at Long CPA	Austin, TX	2442	837,000	23,701	343	9.7	1.5–2.3
Globe-News Center	Amarillo, TX	1300	686,000	19,425	528	14.9	1.5–2.0
Grand Hall at HKU	Hong Kong	920	353,000	9996	384	10.9	0.9–1.5
Richmond CenterStage	Richmond, VA	1800	396,000	11,213	220	6.2	1.6–2.1
Schuster PAC	Dayton, OH	2300	765,869	21,687	333	9.4	1.5–2.0
Symphony Hall	Boston, MA	2625	656,000	18,576	250	7.1	1.9
Tokyo Forum	Tokyo, Japan	1500	604,000	17,103	403	11.4	1.4–1.8
Wagner Noel PAC	Odessa, TX	1800	699,000	19,793	388	11.0	1.6–2.2

Figure 5.1   Table of acoustic volumes for representative large halls.

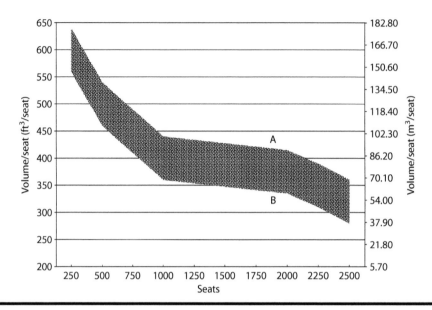

Figure 5.2 **The volume criteria, for acoustic excellence for classical music, varies with hall capacity. (A) Upper limit of recommended volume/seat. (B) Lower limit of recommended volume/seat.**

Information about Volume

- Volume is a rule of thumb, not an exact criteria. Seat size and aisle size are not critical at this point.
- In a proscenium hall, neither the musicians onstage or in the pit are accounted for in the V/seat parameter because the stage and pit volumes are considered separate acoustic volumes.
- In a one-room concert venue with no proscenium, artists and musicians are counted, and there is a higher target V/seat. Typically, a value of 400 ft.³ (11 m³) per person is reasonable for a concert hall.
- As the audience size decreases, the V/seat increases. This is not necessary to keep the RT high but more to keep a high volume to prevent "overloading" of louder instruments.

Achieving an adequate volume for proper RT in multi-use halls is more complex than one might imagine. There are innovative ways to locate volumes in positions above ceilings, behind side walls, and above orchestra shells. Keep in mind that a 2000-seat hall with 375 ft.³ (10.6 m³) per seat has 750,000 ft.³ (21,200 m³) of volume, which is enormous.

Information about Acoustic Volume

- *Attic volumes located above ceiling clouds or acoustic reflectors help the hall be more intimate as the visual space is confined.* The uses of this treatment are described more fully in the case studies section of this book and are highlighted in such halls as the University of Texas–Permian Basin Performing Arts Center and the Schuster Center. Volume placed above the ceiling or attic is in a very effective position because the mean free path is large. Massive roof slabs and walls parallel to each other foster a rich and powerful reverberation. Attic volumes should be concentrated and largest near the front (stage) end of the hall nearest to the sound source. Care must be given to the design of air ducts so that they do not absorb or obstruct the reverberant energy development in the attic.

■ *Volumes behind the hall's side walls and side boxes can add needed volume and create the image of a more compact seating configuration.* This is tricky as the volumes need to be sufficiently large and located near the sound source (stage or pit) so that they are energized and positive contributors to the RT. Small volumes or chambers have little or no value to add RT, and they can actually lower the overall room RT by having a small mean free path.

Examples of Adding Volume

Volumes were added successfully within the Globe-News Center in Amarillo, Texas. The Globe-News volume extends both over and around the sides of the hall in an extreme concrete volume enclosure that is well connected through openings in the heavy wood inner wall and ceiling system (see Figures 5.3 and 5.4).

Volume was successfully added above and around the orchestra shell at Bass Hall in Fort Worth, Texas, and at the Tokyo International Forum in Japan. Designed with the 375 ft.^{3}/person ratio, these additional volumes were not calculated to enhance the hall's overall volumes but rather to add resonant volume to increase the sense of envelopment for audiences near the orchestra and for performers. The Japanese wanted a name for this tool and coined the rather awkward term Concert Hall Shaper (see Figure 5.5).

Exceptions

Volume for reverberation development includes all areas that are acoustically well connected to the main space, including attic spaces and underbalconies, except when those volumes are not

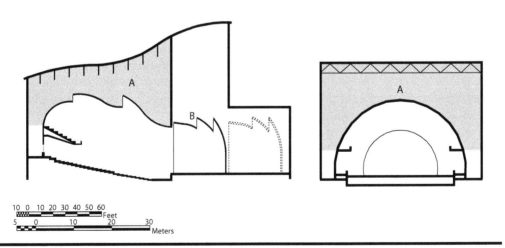

Figure 5.3   Globe-News Center, Amarillo, TX, 2006, 1300 seats. (A) Acoustic volume located above and around the inner walls and ceiling. (B) Orchestra shell moves as a unit and stores upstage in a purpose-built garage.

Figure 5.4 Globe-News Center, Amarillo, TX, 2006, 1300 seats. Inner walls and ceiling are open to the volume above and around through generous openings. Locating volumes here is not for the novice acoustician.

significantly open. Well-connected volumes have openings that allow sound to move freely in and around it and do not significantly absorb sound. For example, the deep underbalcony volume of a historic movie palace is not considered. Neither are air plenums under seats.

An attic volume may or may not be part of the volume depending on its openness and connectedness to the main volume. I err on the side of designing the attic volume very open and with the smallest possible number of clouds or hung ceiling elements in order to foster a well-connected volume, increase mean free path, and reduce absorption.

I was once asked if the RT of a volume could be reduced if the surfaces of a hall were made very hard and sound reflective. Perhaps this architect was thinking about the effect of singing in the shower when volume is small, and RT is large. In fact, the RT is not very large, and the shower is not reverberant; it is resonant. Thus, the answer is no; lining a room with ceramic tile will not mean that the V criteria will be lowered.

Setting the volume for the multi-use hall is a critical starting point and must be checked throughout the design process. The volume must be located in an area where it is supportive of the development of reverberation and where it can temper the loudness of full symphonic ensembles or amplified performances. See Chapter 8 for volume guidelines.

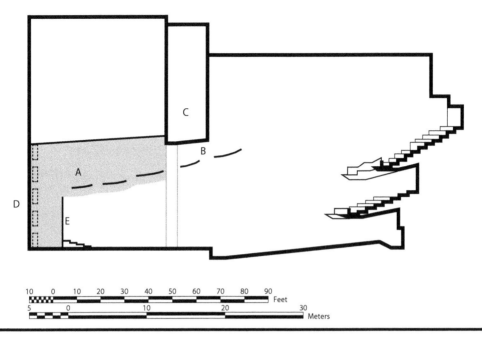

Figure 5.5   Tokyo Forum, Tokyo, Japan, 1997. Hall C, 1500 seats. (A) Concert Hall Shaper volume used around and above the acoustic shell. (B) Motorized forestage ceiling reflectors store in zone C. (D) Shaper ceiling and reflectors store on upstage wall. (E) Shell towers on dollies.

Room Volume

When a stage house exists for opera, ballet, and Broadway-style performances, the volume of a multi-use hall should be measured in front of the proscenium and exclude the stage shell. In order to be effective, the volume high in the room must be open and free-flowing and have minimal partitions or other elements that would reduce the mean free path. Acceptable elements include suspended roof structures, steel supports, and some ductwork (preferably round for minimal low-frequency absorption) because they can actually be effective at adding sound diffusion. Angling the roof line adds volume to the portion of the hall nearest the stage and pit where the sound is generated and helps to energize the volume. A good example of a space that uses this technique is the Wagner Noël Performing Arts Center in Midland, Texas featured in Appendix X. I do not recommend locating these types of volumes far away from the stage or near the rear of the hall (see Figures 5.6 and 5.7).

Examples

At the Long Center's Dell Hall, modest volume zones were added to the cheek walls near the stage. The volume added in the cheek walls allowed the roof to be built slightly lower resulting in cost savings. This volume was open to the main volume through very large and open sound-transparent grilles in the side walls, which also communicated with the attic volume above. As a location for adjustable acoustic curtains, this zone offered the advantages of being close to the stage, being tall enough for drapes to hang unimpeded, and being hidden from view. This is discussed in more detail in Appendix VII (see Figures 5.8 and 5.9).

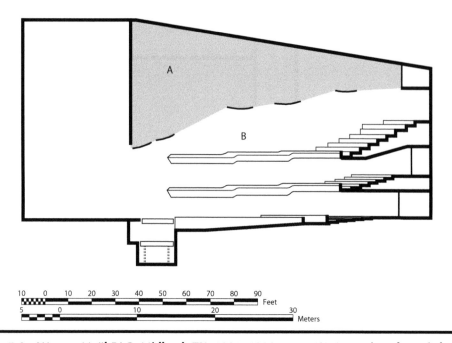

Figure 5.6 Wagner Noël PAC, Midland, TX, 2009, 1800 seats. (A) Acoustic volume is located above the ceiling reflectors here. But unlike the Globe-News Center, it is one volume with the rest of the hall volume, (B).

Figure 5.7 Wagner Noël PAC, Midland, TX, 2009, 1800 seats. The star field of LEDs defines the ceiling but is acoustically transparent.

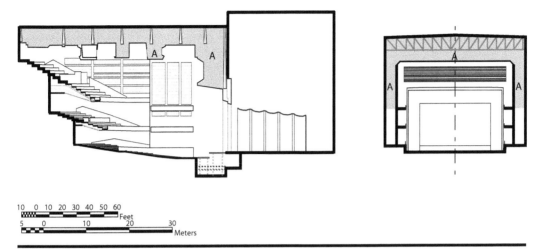

Figure 5.8 Dell Hall at Long CPA, Austin, TX, 2008, 2400-seat multi-use hall. (A) Acoustic volumes above and around the hall provide proper volume/seat ratio while allowing the hall to feel more intimate, and provide early reflection surfaces.

The Globe-News Center in Amarillo, Texas, has a wood inner jewel box with an outer volume container. This acoustically complex design allows inner walls and the ceiling to create early reflection surfaces, increase C80, reduce initial time delay gap, and add diffusion. The outer box creates the large volume needed for extended RT in support of classical ensembles. It was built of massive concrete poured in place to reduce absorption. Large adjustable acoustic panels made of fiberglass are hidden in the upper volume and can be tracked into the volume to reduce RT in support of amplified events (see Figure 5.10). This is covered in Chapter 15.

Figure 5.9 Dell Hall at Long CPA, Austin, TX, 2008, 2400-seat multi-use hall. Sound transparent grilles on sidewalls and in proscenium zone open to the volume behind and are backed by visually opaque scrim materials.

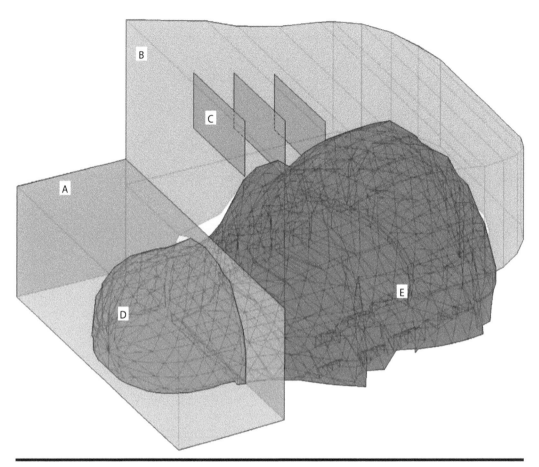

Figure 5.10 Globe-News Center, Amarillo, TX, 2006. CATT model. We use software models where the interaction between the volumes beneath the ceiling and above is complex. (A) Stage volume used to vent high sound pressure levels from inside orchestra shell (D). (B) Attic acoustic volume. (C) Framed acoustic panels control RT. (D) Orchestra shell. (E) Inner walls and ceiling of the hall have strategic openings into the attic volume.

Chapter 6

Chapter 6

Myths and New Halls: Schematic Design through Construction Documents

---

Introduction

After the programming process is complete, the client approves the budget and the design schedule. Then, the acoustician begins the schematic design phase for the new hall. During this phase, it is relatively easy to make changes to the scope because the designs are still malleable. As project design progresses, it becomes more difficult to make changes. Therefore, critical acoustic issues should be identified and brought to the team's attention as soon as possible. The acoustician must proactively advocate for acoustics issues and refine them as the architectural vision of the interior and exterior evolves.

Suppress the urge to be rigid or to reuse acoustic solutions that have been successful elsewhere. The acoustician must look at each design with fresh eyes because no two halls are alike. Each hall has unique locations, user groups, audiences, performers, and clients. Fight the impulse to design in contingency factors with the idea that the client might need to reduce quality to meet a budget. Design what is really needed based on logic, physics, and experience. Use facts and data to support your design.

Acoustic Design Myths

Some acoustic truisms are, in fact, not true. This is my opinion based on many years of research and experience with hundreds of new hall designs. I admit to holding many of these beliefs early in my career before I learned to question, test, and eventually develop new ways to achieve acoustic excellence. The following are examples of widely held acoustic myths.

Myth #1: Multi-Use Halls Should Be Limited in Width

Early in my career, I lobbied to limit seating area width to no more than that of Boston Symphony Hall's 86 ft. if programming included symphonic performance. I understood that this width was vital to providing the proper C80 reflections to create intimacy, clarity, and high loudness levels (see Figure 6.1). Later, I learned that a narrower width in the throat area was critical to providing early reflections for C80 and G but that a narrow rear had no benefit. It actually reduces acoustic quality. See Chapter 12 for more details on hall width.

Myth #2: Halls Should Be as Quiet as Humanly Possible

I listened to halls that were very, very quiet and found the quiet levels to indeed enhance the quality of the performances of unamplified music. I also listened to halls that were almost as quiet (NC-15) and could perceive no qualitative difference in room acoustics or in my ability to hear performers. This myth is related to the data set used to set the human threshold, also known as the Fletcher–Munson curve. Harvey Fletcher and W. A. Munson coined this term to explain how the lowest curve is what they measured as the threshold of human hearing across the entire spectrum. This threshold is based on young adults with perfect hearing, something that very few people have. Our hearing worsens as we age, and our sensitivity is reduced, especially at higher frequencies. Therefore, listeners and performers in halls at NC-15 suffer no degradation of performance or acoustic quality. There is no need to squander

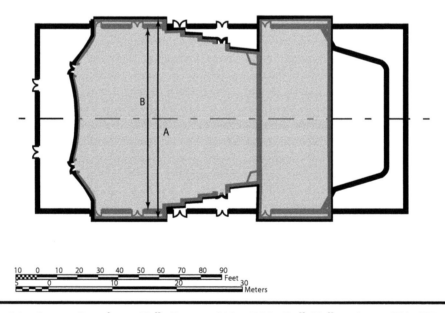

Figure 6.1 Boston Symphony Hall, Boston, MA, 1900; Dell Hall at Long CPA, TX, 2008. Comparison between the 86 ft. (26 m) width of (B) Boston Symphony Hall with the 104 ft. (32 m) Dell Hall (A). It is a myth that halls should be long and narrow, in a shoebox shape, for excellent acoustics.

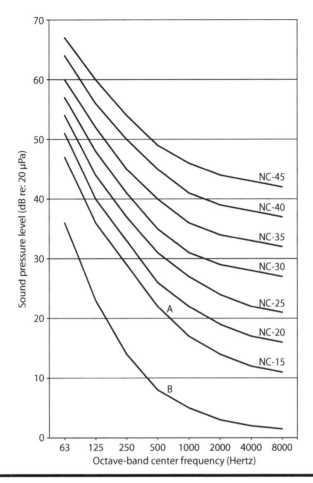

Figure 6.2   Graph of NC levels and threshold of hearing. NC-15 is totally inaudible to audiences. Do not burn budget to achieve theoretical figures. (A) NC curves from NC-15 to NC-45. (B) Approximate threshold of hearing for continuous noise.

resources to reduce that last 15 dB of sound to reach that threshold; very few people can perceive that minute difference (see Figure 6.2).

Myth #3: The Perimeter of the Hall Must Isolate 100% of All Possible Intrusive Sound

I believe this myth to be ludicrous. I once heard an acoustician boast that a new multi-use hall in an urban environment was designed to isolate a bomb blast directly behind the stage house and that a lightning strike on the stage would not be heard during a performance. I consider this to be the epitome of uncontrolled acoustic ego—the acoustician forced the client to take on the vast cost of such unnecessary sound isolation measures. Great halls should be designed to isolate 90% of loud events that might occur. The cost savings for this approach will ensure the economic viability of the facility (see Figure 6.3).

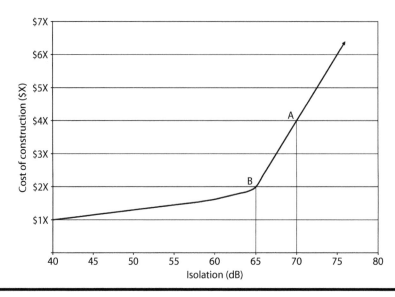

Figure 6.3 Graph of cost of sound isolation. (A) 65 dB may be sufficient to isolate 90% of sound. (B) 70 dB isolates 95% of sound but at a great cost increase.

Myth #4: Seating Under Balconies Must Be Avoided or Limited to One or Two Rows

At first, this seems logical. The lower ceiling overhead blocks the sound that might reach the listener from above, thus reducing the envelopment of surround sound. In actuality, many acoustically successful halls have four-to-six rows under the balcony on the orchestra levels and six-to-seven rows under the first balcony without suffering from ill effects of the overhang. The truth is limiting seating under the balcony works against acoustic quality. For a hall with a reasonable width, more balconies must be added or the length increased in order to accommodate the requisite number of rows. This reduces the G level and intimacy by locating many seats further from the stage.

Myth #5: Adjustable Acoustic Systems Are Not Very Effective

It is true that early drape and moving ceiling systems were not very effective at reducing full-range reverberation for amplified events. Old systems were effective only at controlling mid- and high-frequency energy; however, new and highly efficient systems now exist. These new systems consist of adjustable acoustic banners, moving panels, and low-frequency absorbers. Jaffe Holden helped to develop a system of wool serge panels suspended 4–6 in. (about 155 mm) apart that improved absorption far better than standard velour drapes. The key to adjustable systems is proper location, size, and application. Chapter 15 is devoted to adjustable acoustic systems (see Figure 6.4).

Figure 6.4 AcouStac Banners co-developed with Jaffe Holden store compactly and use wool serge with a defined airspace between layers. (Photographed at Lincoln Center for the Performing Arts in New York City.)

Chapter 7

Myths and Renovations: Transforming Existing Buildings into Multi-Use Halls

Introduction

A renovation makes an acoustician's job harder in some ways yet easier in other ways. At the onset of a renovation, it is essential to have a thorough understanding of the existing building before any decisions are made. It is vital to review acoustic measurements including ambient noise and vibration. Do not spend time measuring HVAC systems if the building will be gutted because those systems are likely to be replaced.

Transportation Noise and Vibration on Site

There are many formal sources that annotate how to take acoustic measurements and provide national and international standards. Therefore, that will not be discussed in this book. It is important to note that measurements involving site vibration from railroad, bus, and automobile traffic are vital in determining if a project is feasible at a reasonable cost. Keep in mind that the best approach is often to isolate the rail systems of light rail and subway rather than try to isolate the existing building. New technologies that employ rubber and neoprene can be mounted to the rails themselves. In complex situations, the entire rail bed, including track and ties, may need to be mounted on a floating inertia base of thick concrete to achieve sufficiently low ground-borne vibration and noise levels in halls (see Figure 7.1).

Aircraft flyover at low altitudes above an existing building may also cause problems. Consider selecting a different location if the facility is on the direct landing path of commercial aircraft because it may not be possible to achieve the mandatory roof isolation to achieve NC-15 in the hall within a reasonable budget.

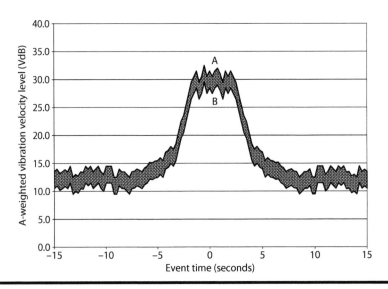

Figure 7.1   Alice Tully Hall, New York, NY, renovated 2009. Vibration measurements from the MTA subway near Alice Tully Hall showed high levels of vibration from express trains (A) and less from the slower local trains (B). Mitigation involved rubber pads under the rails, and floating floors and walls inside the hall. Alice Tully Hall is a featured case study.

Predesign Project Site Inspection

If the structure is solid, the roof is substantial, and the site is expected to be quiet, then it is quite possible that the existing structure can be converted into an acoustically successful hall. The first order of business is to acquire a bright LED flashlight with a focusable beam and a good camera. The first site visit is no place for a cell phone camera or flashlight app.

Plenum

Tunnels or a large air plenum are often located under the seating floor structure of many theaters. Crawling down in these spaces, often full of vermin and 50+ years of moldy popcorn, may be disgusting, but acousticians are compelled to inspect these spaces because they could serve as air supply zones. Repurposing a plenum for HVAC can lead to significant savings. Note the dimensions of the space and any restrictions that might cause excessive velocity noise or turbulence for a new system.

Walls

Thick brick or concrete walls of older buildings provide ideal acoustic reflectors and sound isolators. However, be sure to check the stiffness, thickness, and mounting of historic plaster, wood, or drywall. Finishes may be mounted directly to the brick or masonry or furred off the surface. Taking measurements at an electrical outlet or opening may reveal the actual wall thickness. Pound on the wall with your fists to determine if it is hollow and resonant. If it moves easily or booms loudly, then the materials are thin and not well braced, which could lead to excessive low-frequency absorption. A dull, solid thud when pounded tells the story of a very thick plaster or plaster on masonry finish.

Example: Alice Tully Hall

During an inspection at Alice Tully Hall, Lincoln Center, New York, NY, I pounded all the wood slat walls I could reach and made an educated guess that the walls consisted of 1/2 in. (12.7 mm) thick wood panels. Peering in outlet boxes confirmed my estimate, yet the RT measurements were below this calculation. Staff insisted that the lower and upper walls were the same, and my observations were aligned with staff observations. However, when I used a long ladder to reach the upper regions of the side walls, I discovered that the walls contained glass fiber behind the wood slats. Finding significant areas of absorption on the walls was a shock, but it explained the discrepancy in measurements and supported the hall's reputation for sound dying out midway through the chamber.

Structure

A quality acoustician is obliged to be knowledgeable in many disciplines, including structural forensic engineering. We must be cognizant of steel and concrete structures, column size and composition, beams, trusses, and cross bracing, as well as connections and underpinning foundations. Understanding the basic structure of a building allows acousticians to understand if we will be able to demand thicker and heavier walls, ceilings, and floors and if the volume of the hall can be increased to add reverberation. Wood-structured buildings pose a challenge as they may be of lighter weight and thus more prone to vibration and noise isolation issues that would necessitate expensive fixes to meet NC and RT criteria. Heavy-timbered buildings can actually be wonderful spaces for conversion into a multi-use hall because the thick wood is inherently fire resistant and massive.

Roofs and Attics

Crawling through dark attics is par for the course. The roof is expected to provide isolation from ambient noise and not resonate during heavy rain. Historic roofs tend to be lightweight decking and made of materials such as Tectum, wood, or concrete planks and lightweight precast. Low mass leads to poor noise isolation. If the roof needs to be strengthened, the design team must ensure that the structure can handle the additional weight. Alternatively, the ceiling may be too low to provide adequate RT for the repurposed space. The acoustician can consider removing the existing ceiling and exposing the attic volume in order to increase volume.

Stage Houses

If the existing building has a stage house, the acoustician should make the climb to the stage house grid to investigate. Is there an old smoke hatch that must be replaced to increase roof isolation? What is the roof composition in terms of isolation? Care should be taken as these grids are often 60–80 ft. (18.3–24.5 m) off the stage floor. As a young man, I once climbed a very tall straight ladder to view the grid in an historic theater. Halfway up, the anchors in the soft brick began to pull out, and the old steel ladder began to pull away from the wall. I gingerly but very swiftly descended, sweat rolling down my back. I am much wiser now and caution acousticians to learn from my mistakes. Be careful in these potentially dangerous situations.

Renovation Do's and Don'ts

Renovating halls to become multipurpose performing arts centers is a cost-effective and energy-efficient way to repurpose a building. The case studies within this book explore a few renovations. In the meantime, here are some pitfalls and pointers for the practitioner to consider.

Historic Movie Theaters

Historic movie theaters built in the golden era of movies (1920s to 1940s) are glorious spaces with evocative interiors and sumptuous lobbies. Some can make good multi-use halls if the stage house can be expanded to accommodate a full orchestra, the proscenium can be widened, and adequate volume exists in the hall. Often, none of these things are present. The large, single balcony in many palaces brought viewers close to the screen, but it also divides the acoustic volume into two discrete and disconnected spaces. Mean free path is low; volume is low; and the deep underbalcony is starved of overhead reflections, RT, and envelopment. The acoustician can work on closing off the rear rows of the underbalcony and resloping the orchestra floor to provide better site lines. Leg room can be increased and sight lines improved by resloping and building up the balcony risers with lightweight gauge-framing systems and plywood floors on metal decking. Lifts can be utilized to bring the orchestra past the proscenium and out into the hall, and a storable orchestra shell can be used to surround the orchestra with a blending, projecting, and supportive environment. Concave sound-focusing ceilings can be tamed with flown acoustic clouds that block sound from reaching them. For more information, please see the case study featuring the Carpenter Theatre renovation at the Richmond CenterStage in Richmond, Virginia (see Figure 7.2).

Electronic Architecture Systems

If volume and RT are low, an electronic architecture system can help enhance RT and improve the underbalcony experience for symphonic and opera productions. Modern systems are reliable and mostly invisible. See Chapter 16 for an in-depth discussion of these systems and their applications to renovations and new spaces.

HVAC

It can be a challenge to find an appropriate location for HVAC systems, and engineers are often compelled to place units on the lobby or stage house roof. Do not place HVAC systems on the roof. An NC-15 level was achieved at Ford's Theater in Washington, DC, by placing very quiet air units in an attic space above a plaster ceiling and enclosing them in multilayer drywall enclosures. The ceiling had minimal penetrations and was covered in insulation to increase the transmission loss and lower reverberation in the attic.

Planning for Future Needs

The Fox Theater in Spokane, Washington, is a hall with wall shaping that is very conducive to early reflections. The ceiling was a decorative sunburst shape that served as a forestage reflector, and the narrow width of the proscenium was operative if the orchestra was positioned halfway out in the hall. My team recommended adjustable acoustic drapes to lower RT for amplified events, but that was not possible due to historic preservation guidelines and cost issues. The RT was

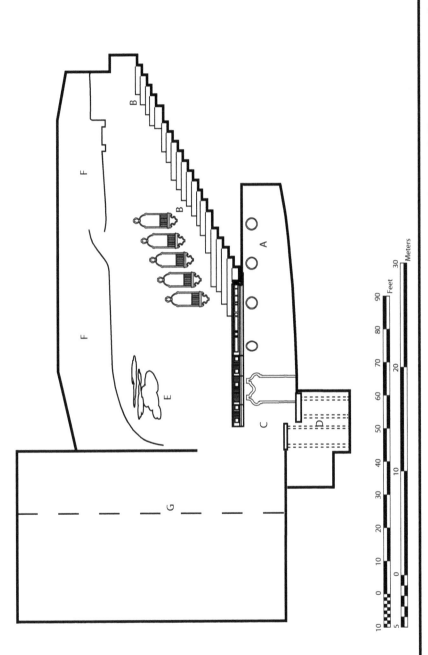

Figure 7.2 Richmond CenterStage, Richmond, VA. This historic movie house renovation faced many challenges. (A) Deep underbalcony on orchestra floor creates poor acoustics for these seats that is hard to mitigate. (B) Old row-to-row spacing creates poor sitting comfort and sightlines. (C) Extending the stage into the hall often requires re-raking the seating on both levels to achieve site lines to the artists. (D) Historic theaters often have very small orchestra pits that are difficult to expand. (E) Ceiling over the forestage is often poorly shaped (concave) and creates acoustic havoc. Flying clouds can help but are complex and expensive. (F) Attics of old theaters have noisy ducts that need replacing to achieve NC-15. (G) Stage houses are often very shallow and too small for a full symphony and chorus. This figure shows the expanded stage at Richmond CenterStage, also featured as a case study.

Figure 7.3 Martin Woldson Theater at the Fox, Spokane, WA, renovated 2007. This theater once hosted Paderewski, Vladimir Horowitz, and Frank Sinatra, and is now home of the Spokane Symphony. Features include a new orchestra shell, stage extension, and a rebuilt balcony.

calculated to be about 1.6 seconds when the hall was occupied, slightly too dry for the Spokane Symphony. Space was set aside for future racks and a small amount of conduit specified for a future electronic architecture system if needed. The modest RT was offset by excellent C80 and NC-15 noise levels and by the extraordinarily beautiful renovated building, and the electronic architecture system has not, to this date, been installed. A wood-veneered Diva orchestra shell manufactured by Wenger completes the stage and can be easily stored offstage for other events (see Figure 7.3).

Overcoming Negative Attributes

The renovation of Baldwin Auditorium at Duke University in North Carolina involved every aspect of acoustic renovation possible and achieved outstanding success. Our first inspection

Figure 7.4 Baldwin Auditorium at Duke University, Durham, NC, 1930. The hall before renovation was an acoustic nightmare. It was overly reverberant, had poor early reflection patterns and had an awful shell. Baldwin Auditorium is featured as a case study.

yielded many negative surprises. Mechanical units located in the large air plenum under the seats of the hall were so noisy that they had to be shut off during performances. All the walls and ceilings were covered with thin mineral acoustic tile glued to plaster, and uncushioned seats made entirely from wood created an odd frequency–selective RT. The hall was wider than deep, which necessitated narrowing the hall to improve C80 reflections, and the stage behind the proscenium was stuffed with ancient acoustic shell towers and drapes. The faculty despaired that nothing could be done to improve the acoustics especially since the space featured a large sound-focusing dome above the seating area.

After conducting a survey, I discovered positive attributes that had been previously overlooked.

A beautiful interior existed under the caked-on mineral acoustic tile. It featured hard solid plaster, a good sign for excellent RT and BR. Each bay within the dome was also covered in mineral tile. This could be diffused to add RT and improve diffusion overall. Reflections from the dome could be blocked by a simple forestage reflector system, effectively forcing sound into the audience. If HVAC units could be relocated, the large plenum could be a huge benefit for duct routing and a noise reduction buffer space. The large attic space could provide a buffer for exterior noise. It occurred to me that a reduction in seat count could offset many existing acoustic challenges. The hall could be narrowed to improve C80, a wraparound balcony could be added to increase intimacy, and the stage could be expanded into the hall to enhance sound for large ensembles. Lastly, I observed that new interior wall systems could mitigate sound leakage through old side wall windows (see Figure 7.4). See the case study on Baldwin Auditorium (Appendix II) for more details on this renovation.

Myths and Misconceptions

Myth #1: Older Halls Are Always Acoustically Better than New Halls

There are actually many new halls that have exceptional acoustics. Many older halls with poor acoustics were demolished to make way for new buildings, burned to the ground and never rebuilt, or destroyed in war. For the most part, only the great halls remain.

Myth #2: Aged and Seasoned Wood Has Better Acoustic Qualities than New Materials

This is not true. Woods can dry out, split, and crack causing additional sound absorption and reduction in high-frequency RT over time. This happened to the thin wood panels and slats at Alice Tully Hall at Lincoln Center in New York City. It is true that lumber used to be thicker (a 2 × 4 was really 2 × 4 in.) but that thickness difference is of no acoustic consequence. Wood stage floors get worn down by years of rolling pianos, stage risers, and heavy crates and must eventually be replaced. Wood stage floors at Boston Symphony Hall, Carnegie Hall, and Musikverein have all been recently replaced with no acoustic degradation.

Myth #3: Halls Mellow and Age, and the Acoustics Improve Over Time

I have heard this comment more than a few times from critics and musicians. There is no evidence that this is true after the first year or so. Early on, the hall's materials adjust to the humidity and temperature in the hall. Wood, plaster, and concrete continue to cure and dry after being installed. After a year though, the moisture content has stabilized, and few changes will occur.

Myth #4: Carpet Is an Acoustic Material

Older theaters often have excessive carpeting, which absorbs too much of the high frequencies and overtones and unbalances musical timbre. Removing carpet often improves acoustics for most performances. In fact, carpet can be useful in some locations and for some purposes. Use a thin carpet with a low pile height sparingly in aisles to eliminate footfall noise from late patrons.

Myth #5: Old Plaster in Halls Is Better than Newer Plaster

This may be only partially true. Plaster on metal or wood lath was applied by craftsmen years ago in naturally varying thicknesses. Plaster was a multilayer application, and the first layer squished through the lathing unevenly and yielded uneven thickness. This makes for advantageous non-homogeneous panel resonances, which are spread out across the frequency spectrum. Similar effects can be replicated in today's construction by varying plaster thicknesses and drywall or, more commonly, applying varying thickness diffusion elements to walls and ceilings. See the case study on Wagner Noël Performing Arts Center (Appendix X) for a discussion on drywall walls of varying thickness.

Chapter 8

Running the Acoustic Model

Introduction

Regardless of whether the design is for a new hall or the renovation of an existing facility, it is necessary to run an acoustic model. This important step helps move the concept to reality.

The proper reverberation time (RT) must be calculated early in the design process long before the architects have completed plans. This early RT check is based on preliminary plans and is a prerequisite to the setting of ceiling and roof heights, also determined early in the process. A detailed understanding of the derivation and assumptions used in the Sabine equation is essential to determining accurate RT calculations. Refer to Chapter 2 for a detailed discussion of its derivation and use. Sabine's formula is indispensable in determining a high estimate for the RT. In the real world, the RT is almost always 10%–15% lower than calculation would indicate due to unforeseen absorption from duct openings, floor grilles, and potential contractor's errors in construction.

Volume Criteria

Halls with proscenium openings and 800–2400 seats should have volumes of 325–350 ft.^{3} (9.2–9.9 m^3) per person. Halls in the 350–750 seat range should be 400–450 ft.^{3} (11.3–12.7 m^3). Halls with fewer than 350 seats should be closer to 500 ft.^{3} (14.1 m^3). Note that the volume must be increased to keep the hall from being overloaded by large orchestras and amplified music when there is a small audience capacity.

RT Calculations

Many excellent acoustic modeling programs are commercially available to calculate RT and other acoustic parameters. However, these programs are not an option early in the design process because the design is not yet set. A spreadsheet-based system can help the acoustician to conduct an RT check in the schematic design phase (see Figure 8.1).

Surface/code	Area (in Square Feet)/material	63 Hz	125 Hz	250 Hz	500 Hz	1 kHz	2 kHz	4 kHz
Seating	4000							
	Occupied, lightly upholstered (Beranek 1998)	0.36	0.51	0.64	0.75	0.80	0.82	0.83
		1428	2040	2560	3000	3200	3280	3320
Aisles	2300							
	Nylon thin pile on concrete	0.02	0.04	0.05	0.10	0.20	0.45	0.65
		46	92	115	230	460	1035	1495
Ceiling	7122							
	Plaster on air space	0.25	0.20	0.15	0.15	0.10	0.05	0.05
		1781	1424	1068	1068	712	356	356
Dome	3040							
	Plaster on air space	0.25	0.20	0.15	0.10	0.05	0.04	0.05
		760	608	456	304	152	122	152
Ceiling treatment	760							
	1 in. BAD RPG	0.10	0.18	0.32	0.80	0.91	0.71	0.52
		76	137	243	608	692	540	395
Front wall	922							
	Smooth plaster on block BERANEK	0.15	0.12	0.09	0.07	0.05	0.05	0.04
		138	111	83	65	46	46	37
Drape	1078							
	Velour, 18 oz/sq yd, 4 in. from wall (100% fullness)	0.04	0.06	0.27	0.44	0.50	0.40	0.35
		43	65	291	474	539	431	377
Side walls	4070							
	Smooth plaster on block BERANEK	0.15	0.12	0.09	0.07	0.05	0.05	0.04
		611	488	366	285	204	204	163
Side wall treatment #1	1930							
	6v12 Banners acouStac	0.44	0.81	1.25	0.96	0.94	0.94	0.94
		849	1563	2413	1853	1814	1814	1814
Rear wall	1530							
	Smooth plaster on block BERANEK	0.15	0.12	0.09	0.07	0.05	0.05	0.04
		230	184	138	107	77	77	61
Rear wall treatment	470							
	6v12 Banners acouStac	0.44	0.81	1.25	0.96	0.94	0.94	0.94
		207	381	588	451	442	442	442
Air absorption	271							
	Air (per 1000 ft.³) - relative humidity 40%	0	0	0	0	1	4	10
		0	0	0	0	271	1084	2710
Total absorp. (sabins) 27,023		6168	7093	8321	8445	8608	9430	11322

Volume (in ft.³)	271,000	Total surface area (in Square Feet)	27023

	63 Hz	125 Hz	250 Hz	500 Hz	1 kHz	2 kHz	4 kHz
Reverberation Time – Sabine Equation (in seconds)	2.15	1.87	1.60	1.57	1.54	1.41	1.17
	63 Hz	125 Hz	250 Hz	500 Hz	1 kHz	2 kHz	4 kHz
Reverberation Time – Eyring Equation (in seconds)	1.90	1.62	1.34	1.31	1.28	1.15	0.91
Average alpha	0.23	0.26	0.31	0.31	0.32	0.35	0.42

Figure 8.1 Baldwin Auditorium at Duke University, Durham, NC, 2013. This is an example of a reverberation time worksheet from Baldwin Auditorium that utilizes real-world coefficients.

A consistent measurement technique for volume and area is an essential part of making proper calculations. Confirmed and accurate octave-by-octave absorption coefficients must be obtained for each material. This is more complex than it appears at the onset. The coefficients of plaster and drywall are unreliable at low frequencies, and materials such as carpet can have vastly different absorption rates due to differences in pile height, weight, and backing.

The RT calculation should be updated throughout the design process. It is essential that the acoustician refine the areas, volumes, and coefficients phase by phase so that there are no surprises.

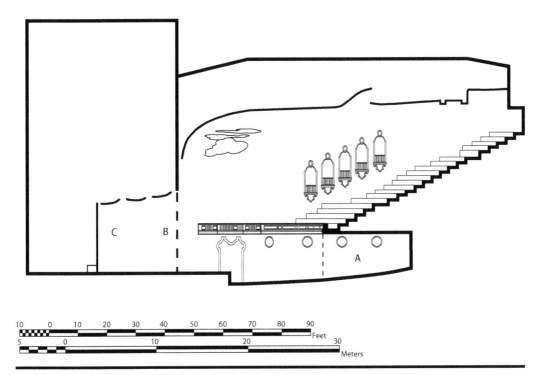

Figure 8.2 Volume calculations. (A) When confronted with a deep underbalcony, do not include the volume. Instead, assign coefficients to the opening. (B) Apply coefficients to the proscenium. Do not include the stage house volume in the calculation. (C) Add the volume of the orchestra and surfaces inside the shell. Remember to add the absorption of the musicians.

Volume

In a closed ceiling room design, volume is easily defined at the ceiling. However, in an open ceiling system, the volume must be calculated to the underside of the roof deck minus any large HVAC ducts. Ceiling types are discussed in detail in Chapter 13.

Underbalcony Volume

Inevitably, the issue of underbalcony volume must be tackled. Include volumes under the balcony only if the area follows the guidelines for good hall design as discussed in Chapter 11 and if the area does not have an excessive height.

Volume is not included in the total when there is a deep underbalcony, such as in a historic movie palace with more than seven or eight rows. The opening into the underbalcony zone at the balcony front is considered a surface with absorption equal to the proscenium opening coefficients (see Figure 8.2).

Orchestra Shell Volume

Orchestra shell volume must be included in calculations for symphonic modes. Absorption by the orchestra shell and the musicians must be included in the calculations.

In general, the volume calculation does not need to be totally accurate early in the process. Later, as the design is finalized, accuracy becomes vital.

Absorption Coefficients

Figure 8.3 highlights a few typical absorption coefficients for calculations. These have been gleaned from a number of texts on the subject, manufacturer's data, and my own measurements. This is only a sampling of a few regularly used coefficients.

Description	63 Hz	125 Hz	250 Hz	500 Hz	1 kHz	2 kHz	4 kHz
ACOUSTIC PLASTER							
Pyrok: 1/2 in. (12.5-mm) Acoustement Plaster 20 with textured finish	0.04	0.05	0.19	0.60	0.95	0.94	0.88
BASWAphon Acoustical Finish System - 2.68 in. (68-mm) system	0.34	0.49	0.75	0.76	0.65	0.62	0.48
AUDIENCE/SEAT ABSORPTION							
Occupied, heavily upholstered (Beranek 1998)	0.50	0.72	0.80	0.86	0.89	0.90	0.90
Occupied, medium upholstered (Beranek 1998)	0.43	0.62	0.72	0.80	0.83	0.84	0.85
Occupied, lightly upholstered (Beranek 1998)	0.36	0.51	0.64	0.75	0.80	0.82	0.83
Unoccupied, heavily upholstered (Beranek 1998)	0.49	0.70	0.76	0.81	0.84	0.84	0.81
Unoccupied, medium upholstered (Beranek 1998)	0.38	0.54	0.62	0.68	0.70	0.68	0.66
Unoccupied, lightly upholstered (Beranek 1998)	0.25	0.36	0.47	0.57	0.62	0.62	0.60
CARPETS							
Nylon thin pile on concrete	0.02	0.04	0.05	0.10	0.20	0.45	0.65
COMMON MATERIALS							
Concrete block, coarse and unpainted	0.20	0.36	0.44	0.31	0.29	0.39	0.25
Concrete block, painted	0.05	0.10	0.05	0.06	0.07	0.09	0.08
Construction concrete, tooled stone, or granolithic	0.05	0.02	0.02	0.02	0.04	0.05	0.05
DRAPERY							
Acoustac 3W12 Banners (6 in. (150-mm) airspace to test chamber wall, closed sides)	0.14	0.77	1.20	0.94	1.00	0.90	0.91
Acoustac 6W12 Banners (6 in. (150-mm) airspace to test chamber wall, closed sides)	0.44	0.81	1.25	0.96	0.94	0.94	0.94
Acoustac 6w11 Banners Hung in space	0.05	0.57	0.91	1.14	1.34	1.51	1.60
FIBERGLASS							
1" (25mm) Owens Corning 705, plain faced, Mounting A	0.01	0.02	0.27	0.63	0.85	0.93	0.95
2" (50mm) Owens Corning 705, plain faced, Mounting A	0.08	0.16	0.71	1.02	1.01	0.99	0.99
3" (75mm) Owens Corning 705, plain faced, Mounting A	0.25	0.54	1.12	1.23	1.07	1.01	1.05
4" (100mm) Owens Corning 705, plain faced, Mounting A	0.35	0.75	1.19	1.17	1.05	0.97	0.98
FLOORINGS							
Cork, rubber, linoleum or asphalt tiles on concrete	0.01	0.02	0.03	0.03	0.03	0.03	0.02
Hard floor tiles, or composition floor	0.02	0.03	0.03	0.03	0.04	0.05	0.05
Wooden Platform with air space below (Beranek 1998)	0.50	0.40	0.30	0.20	0.17	0.15	0.10
GLASS							
Heavy plate glass-1/2 in. (12.5-mm)	0.20	0.18	0.06	0.04	0.03	0.02	0.02
Typical window glass-1/8 in. (3-mm)	0.40	0.35	0.25	0.18	0.12	0.07	0.04
GYPSUM BOARD CONSTRUCTIONS*							
1 layer gypsum board, 5/8 in. (16 mm) thick	0.40	0.30	0.10	0.07	0.06	0.08	0.09
2 layers gypsum board, 5/8 in. (16 mm) thick	0.20	0.15	0.10	0.07	0.06	0.08	0.09
3 layers gypsum board, 5/8 in. (16 mm) thick	0.15	0.10	0.08	0.07	0.06	0.08	0.09
PLASTER CONSTRUCTIONS							
Plaster on lath, smooth	0.02	0.14	0.10	0.06	0.04	0.04	0.03
Plaster suspended ceiling with large air space above	0.25	0.20	0.15	0.10	0.05	0.04	0.05

*on 3-5/8 in. (90-mm) metal studs, 16 in. (400-mm) o.c., 1 in. (25-mm) batt insulation-EST.

Figure 8.3 This is a sample chart in English units showing typical absorption coefficients used in calculations for multi-use halls.

Seat absorption coefficients are based on seat fabric, upholstery, and foam thickness. It is necessary to use the appropriate coefficients in order to obtain accurate calculations. We use Beranek's (2004) indispensable seating coefficients for medium and heavily upholstered chairs as published in *Concert Halls and Opera Houses* to complement our experience and evidence-based design (Beranek 2003). This process is described in detail in Chapter 14.

Seating Zone

According to Beranek (2003), a zone extending 1 m beyond the perimeter of the seats should be included in the seating absorption area to account for edge absorption and to obtain an accurate value for seating absorption. Keep in mind that in symphonic mode, seating absorption is the only significant absorption in determining the final RT.

The absorption of one, two, and three layers of drywall at 63- and 125-Hz bands can only be estimated because of variances in field conditions, framing, and stud spacing. Be conservative and assume that there will be a higher level of low-end absorption than indicated in calculations. In fact, any absorption data at 63 Hz are suspect because most acoustic laboratories are not certified down to 63 Hz.

It is acoustically dangerous to install carpet under the seating if symphonic or opera shows are included in the hall's program. Carpet in aisles is acceptable if it is thin and glued down without padding. As a rule, the less carpet used in a multi-use hall, the better.

When there is no orchestra shell in place, use the proscenium opening estimated coefficients ranging from 0.2 at 63 Hz to 0.45 at 4 kHz. Derivation of this number is based on evidence and reverse calculations after RT field measurements. Use absorption coefficients of the orchestra shell walls, ceilings, and stage floor when in orchestral mode.

Air Absorption

Air absorption is a serious factor especially at higher altitudes or with high humidity. Dry air can significantly alter absorption in large halls and can yield unfavorable results. There are HVAC techniques that can be utilized to increase humidity in the hall and stage in order to achieve predicted RT results.

It is a challenge to calculate the absorption of elements in the attic space of an open ceiling hall. Forestage reflectors floating within the volume are not identical in absorption to thin panels over a contained air space, as listed in the coefficients. This is described in Chapter 13. However, they will display some important mid- and low-frequency absorption. Drape pockets, large sheet metal ducts, and plenums also represent absorption and should be estimated early on.

Drapes and Banners

Acoustic drapes or banners will vary in absorption when deployed at low frequencies based on how they are mounted. Try to target low-frequency reverberation in the 1.3–1.5-second range. This may require as much as 5000–7000 ft.^{2} (141.5–198.2 m^3) of material. Chapter 15 describes adjustable acoustic systems in more detail.

Myths and Misconceptions

Myth #1: Acoustical Modeling Software Predicts Performance

Not so. Models are useful tools that provide good estimates of the acoustic performance of a hall. However, they are not totally accurate especially when it comes to nontraditionally shaped rooms, or rooms with suspended acoustic reflectors, and connected volumes. Models are limited in the number of surfaces that can be calculated, so round spaces are challenging to model. Models depend on the accuracy of the coefficients utilized, and I have found laboratory tests to be unreliable in the real world especially below 125 Hz.

Myth #2: RT Models Only Need to Be Run When the Room Is Fully Designed

This is not true. Even when the room is just a concept, RT models are useful in confirming the appropriateness of the design's direction. If the RT is too low in concept design phase, it will only get lower as the design progresses and the hall volume shrinks with the addition of ductwork, catwalks, circulation, and structure.

ARCHITECTURAL DETAILS

Chapter 9

Orchestra Pit

Introduction

There is very little published on practical solutions for designing acoustics for an orchestra pit. This chapter guides the reader through a brief history of orchestra pits and how to design them for the modern multi-use hall.

Musicians generally abhor playing in a pit. They feel that it is crowded, hot, loud, and a horrible environment in which to hear themselves and other musicians. The acoustician's goal is to improve the acoustic environment within the pit and find the optimal projection of music to the audience and stage.

Early Pits

Orchestra pits in ancient theaters began as a little more than a recessed zone in front of the stage used to locate musicians accompanying singers on stage. Beginning in the eighteenth century, orchestra pits for European theaters and opera houses evolved and became more sophisticated with the addition of movable floor sections and elaborate rails to define the pit area. The pit's depth and size depended on the size of the ensemble and ranged from accommodating 30 musicians to a full symphonic ensemble of 100 or more musicians. From the beginning, it was difficult to balance sound levels between the pit orchestra and the singers.

Open Pit versus Covered Pit

Richard Wagner challenged the notion and design of the traditional pit with his Festspielhaus built in 1876. His vision of the pit orchestra was to make the musicians invisible to the audience and to temper their sound by recessing them deep under the stage edge in a crowded, dark cavern. He achieved the desired mysterious sound for opera orchestras and made musicians invisible by covering them with a metal hood, but he also made both conducting and playing in the pit nearly impossible (see Figure 9.1).

By the 1960s and 1970s, multi-use halls commonly included a pit that extended partially under the stage like a modified version of Wagner's Festspielhaus. This minimized the pit opening and the

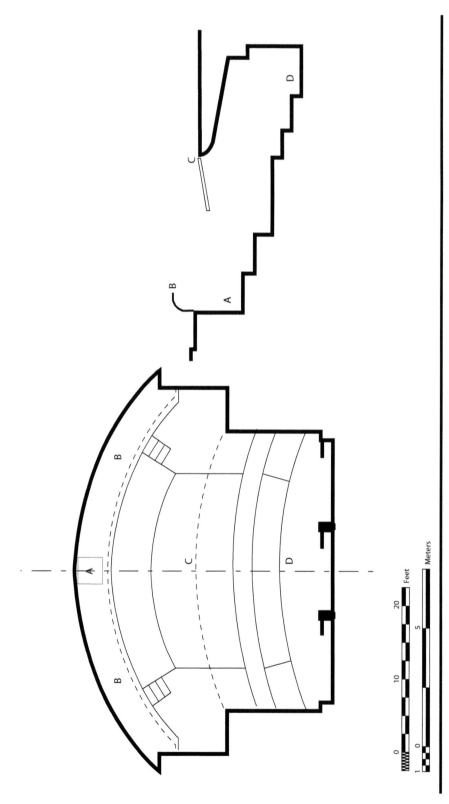

Figure 9.1 Bayreuth Festspielhaus, Bayreuth, Germany, 1876. This early orchestra pit buries the orchestra deep under the stage overhang. (A) Conductor's podium. (B) Pit rail is actually a metal shield that blocks sound from reaching audience. (C) Stage edge with metal shield. (D) Pit rear six levels down.

gap between the stage performers and the audience. This is a good thing for audiences but not for musicians. The sound is contained and reflected back by the low ceiling of the stage overhang. This may not be a problem for the clarinet, but brass, percussion, and piano produce high sound levels that really cannot be made softer without changing the tone and character of the sound. Wagner's famous operas functioned well and continue to be performed successfully today in the Festspielhaus with its cave-like pit and hood cover. However, today's conductors demand an open pit where sound can freely travel into the hall, free of dangerous loudness levels and uncomfortably crowded conditions. Replicating the "mystic gulf," Wagner's term for the detachment of the orchestra from the audience, in a multi-use hall is too limiting to the broad range of programming in a modern hall.

Modern Pits

The recommended approach for pit construction is to open it to expose musicians to the hall and get most musicians out from under the overhang. A modest zone of about 6–8 ft. (1.8–2.4 m) deep under the stage allows for overflow of musicians when the orchestral forces are large. Another zone exists for circulation and storage. Platforms in the pit or lifts can be reconfigured so that more of the high-energy instruments are out in the hall, directly radiating toward the ceiling. The percentage of open to cover should be 75/25 or 80/20 so that most of the musicians are out in the hall. Design the entire pit area on the lifts, and ensure that the area under the stage is large enough. Plan for 17–20 ft.^{2} (1.6–1.9 m^2) per musician. Use the lower number when the pit is to accommodate more than 50 musicians and the higher number for smaller pits.

Broadway Pits

Broadway pits with multiple electronic instruments, large percussion rigs, and drum sets require even more area per musician. Size the pit for no less than 25 musicians for Broadway productions to allow for enough space even if there are far fewer in the band. Draft a pit layout showing the positions of the musicians, and remember to include storage areas for pit stands, chairs, and risers.

Pit Dimensions

Pit dimensions can make or break the pit.

Width

Extend the pit width (stage left and right) as wide as the building allows in order to maximize capacity and reduce the upstage and downstage depth. There is no reason to stop the pit opening short of the side walls or side apron area of the hall. The area under the stage should be as wide as possible, extending out to the proscenium structural columns. This was done successfully Dell Hall at Long Center for the Performing Arts in Austin, Texas (see Figures 9.2 and 9.3). Be careful to avoid parallel pit walls.

Depth

Depending on the pit capacity, the zone under the stage should be at least 5 ft. (1.5 m) deep and might be as large as 14 ft. (4.3 m). The typical range is 6–8 ft. (1.8–2.4 m). Pit lifts are usually

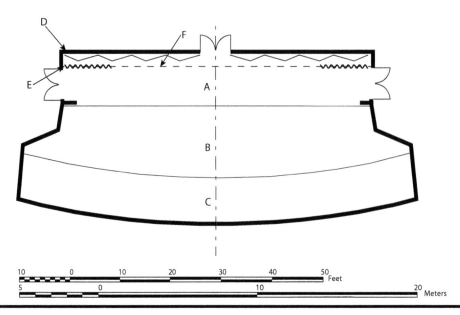

Figure 9.2 Dell Hall at Long CPA, Austin, TX, 2008. Double lift pit. Two lifts provide high level of flexibility. (A) Area under the stage limited to 6–8 ft. (1.8–2.5 m) deep. (B) Lift 1, the main pit lift, extends left and right as far as possible to maximize area. (C) Lift 2 meets the needs of a large opera and ballet orchestras, up to 90 musicians. Note the width left and right. (D) Diffusive upstage wall. (E) Acoustic velour drape or acoustic panels on (F) track.

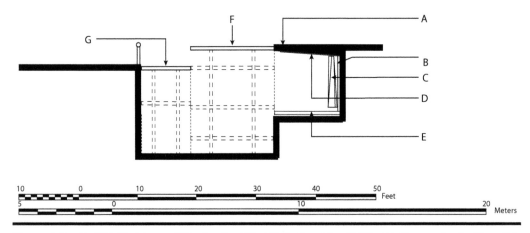

Figure 9.3 Dell Hall at Long CPA, Austin, TX, 2008. Double pit lift. (A) Stage edge held as this as possible. (B) Adjustable acoustic drape or panels on upstage wall. (C) Acoustic panels on ceiling to control loudness. (D) Diffusive wall. (E) Wood floor similar to stage construction. (F) Lift 1 can stop at stage, seating, pit, or storage levels. (G) Lift 2 might only go from seating to pit level in some cases. Lift 2 rises to stage level for orchestra thrust design.

Figure 9.4 Dell Hall at Long CPA, Austin, TX, 2008. First lift is partially complete in the construction photo. Note the Gala Systems Spiralift for vertical movement and steel frame to support wood lift floor.

in increments akin to audience seating row depths such as 8–9 ft. (2.4–2.7 m) for two rows or 12–14 ft. (3.7–4.3 m) for three rows. A double lift, which was used at the Long Center, allows the pit area to be expanded and results in a reduction to the audience seat count (see Figures 9.3 and 9.4).

Height

Vertical dimensions, to the millimeter, are critical to the pit's acoustic function. The pit section shown in the Long Center drawing (Figure 9.3) illustrates these issues. It is critical to minimize the thickness of the orchestra pit lip, also known as the stage edge at the pit overhang, to the absolute minimum dimension. A thick lip forces the floor of the pit down to achieve the code for head clearances. It also throttles the sound from leaving the pit. The target thickness is 12 in. (305 mm).

The conductor must see the feet of dancers on the stage and therefore requires a high podium. An overly thick pit lip thickness forces the conductor's podium to be even taller so that the opening to the pit underside is not throttled. A 36 in. (910 mm) high conductor's podium is the highest recommended for comfortable sight lines from the musicians closest to the conductor.

Pit lifts are movable, motorized platforms that can be stopped at any level, ranging from low storage to above stage floor. The exact location should be determined by the music director during rehearsals. My settings within the text are merely suggestions.

Portable pit platforms, or pit risers, can be used to improve site lines to the conductor. The first few stands of violins and cellos are raised 12–18 in. (300–460 mm) so that they can view the conductor without craning their necks. For the convenience of musicians, platforms can also be used to raise the pit floor under the stage overhang to the lift height.

Acoustic Treatments

There are many acoustic treatments that can be applied to control sound within the orchestra pit. These include ceiling treatments, wall treatments, and wall shaping.

Ceiling Panels

Treat the ceiling of any overhang with the acoustic panel's surface mounted to the concrete structure using 1 in. (25 mm) thick acoustic panels rated Noise Reduction Coefficient (NRC) 0.9. Full, 100% velour acoustic drapes should be positioned on the upstage wall in 8 ft. (2.4 m) wide panels to temper the loudest levels. Cover the entire wall with the drapes in 8 ft. (2.4 m) wide panels, not a continuous drape with a pull cord. Users will then be able to space the drapes and gather them to maximize efficiency. The wall behind the drape should be gently articulated for modest diffusion and made of one or two layers of drywall.

Wall Panels

As an alternative to the velour drapes, movable panels on upstage walls can be reversed to be either absorptive or diffusive. I have seen walls lined with fine wood veneers and paneled to resemble a library or drawing room. I am not convinced that this technique produces any discernable improved acoustic performance over the drywall and acoustic drapes described previously, but it may improve the psychoacoustic performance.

Pits exclusively used for Broadway and amplified productions require more elaborate acoustic treatments. Take the Hobby Center for the Performing Arts in Houston, Texas, for example, where Sarofim Hall's pit is used exclusively for amplified music. Here, my firm created a sound recording studio environment and covered all the walls in 4-in. (100-mm) acoustic panels. The pit floor under the stage was a series of movable platforms instead of a permanent structure, thus providing complete flexibility.

Wall Shaping

A curved wall at the downstage seating area should be multilayer plywood, drywall, or a similar smooth curved material. This ensures that there is no gap between the lift and the wall where seating wagons are stored.

Pit Rail

The pit rail is the area between the pit opening and the audience. Some acousticians recommend the pit rail to be solid, and others say that it should be sound transparent to allow sound out. Depending on the music and the production, both opinions are correct. If the goal is to project natural sound out of the pit, then an open rail with thin, light-blocking fabric cover works well.

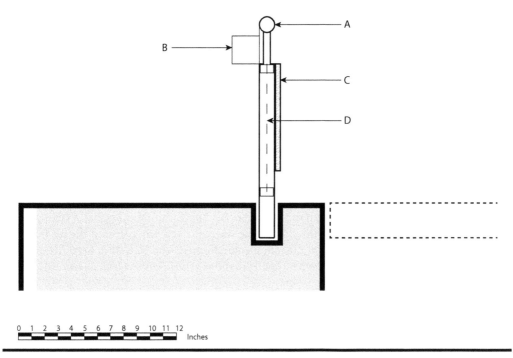

0 1 2 3 4 5 6 7 8 9 10 11 12
Inches

Figure 9.5   Pit rail. Ballet orchestras prefer a pit rail that is sound transparent. For other uses, the pit rail should be solid and sound reflective. (A) The railing is movable and should be light weight for ease of use. A simple aluminum railing is best. (B) Rail must allow for speakers and lighting instruments, and perhaps a safety net. (C) Fixed or removable sound reflector on inside of rail. (D) Metal mesh or fabric to block light from pit when reflector is removed for ballet.

This might be the case during a ballet or dance performance or when period instruments with low sound power levels are being used (see Figure 9.5).

It can be tricky to balance singers and musicians in the pit. A solid rail tends to keep sound levels low as they come out of the pit and in better balance with singers. Unless there are top-tier opera voices, pit musicians usually need to hold back to stay in balance, and a solid rail helps. Even a section of 1/2 in. (13 mm) thick plywood makes a difference.

Some pits have removable pieces of solid wood on the rail so that the pit can be tuned to be open or closed. It is important to remember that the large opening at the top of the pit will leak significant levels of sound into the upper reaches of the hall. Realistically, the rail treatment can only affect the orchestra seats within 8–10 rows of the pit.

Pit Floor

The pit floor lift area can be seated with audience chairs or raised to stage level. The floor surface must match the stage floor. The airspace under the wood floor mounted on the steel lift frame provides an airspace of resonant support for pegged instruments and piano. Pegged instruments should be planted on a wood surface, and, therefore, the stage's wood floor should be replicated in the area under the stage. Never carpet a pit. A small rug or heavy carpet section located under the percussionist can make a difference in loudness and improve balance, but that is all the carpeting that is recommended.

Pit Ventilation

Let us face it; the pit can be a horrible environment for musicians to create art. It is dark, out of view from the audience, cramped, and loud. The pit often has an uncomfortable air temperature. Areas under the overhang get hot and stuffy, whereas open areas get chilled from the hall's air-conditioning systems.

Providing supply air grilles at the upstage rear of the pit is a modest-cost approach to quietly ventilate the pit. Take care in coordinating the ducts with the acoustic drapes, the diffusion surfaces, fire sprinklers, and lighting. Supply air slowly at the floor for minimal conflicts and drafts.

At Marion Oliver McCaw Hall in Seattle, Washington, Jaffe Holden perforated the pit floor and provided fresh conditioned air from below via thousands of tiny openings. That technique proved very successful. Adding a dedicated pit unit and new ductwork at the Wortham Theater Center in Houston, Texas, had mixed results because the main air ducts in the hall's ceiling could not be changed. Cold drafts pushed through, while warm air was pumped into the pit from the sides. At the David H. Koch Theater at Lincoln Center, formerly the New York State Theater, the pit lift floor was perforated similarly to McCaw Hall; however, the air was exhausted rather than supplied (see Figure 9.6).

Figure 9.6   McCaw Hall, Seattle, WA, renovation 2003. Pit ventilation using perforations. Holes drilled in the floor allow for supply or return air to condition air in large orchestra pits.

The Metropolitan Opera House at Lincoln Center for the Performing Arts as Benchmark

The orchestra pit will become the focus of the acoustic design when the multi-use hall is to be home to an important opera or ballet company. I find it useful in programming and design reviews with opera and ballet companies (and the design team) to reference well-known national and international opera house pits for height, width, and depth dimensions.

Misconceptions are common regarding the size of a typical lift platform, the area under the stage overhang, and the height of the pit floor relative to the stage. In an effort to provide a benchmark for the reader, details on arguably the most important orchestra pit in North America, the Metropolitan Opera House at Lincoln Center for the Performing Arts (the Met), are included in this chapter (see Figures 9.7 and 9.8).

In the United States, the pit that many musicians admire and prefer for excellent acoustics is the orchestra pit at the Met. I have had the good fortune of being an advisor to the Met regarding pit modifications over the years. For the reader's reference, we have included plans and sections of the Met pit's base layout as it was in the 2014–2015 season. The floor platforms and risers can be modified and adjusted to accommodate the needs of every opera production within the season. In the Mozart opera configuration, only the upstage lift is used for musicians. Over the years, each maestro's preferred pit lift elevations were recorded not in the lift mechanisms' controls but rather in white paint on the side walls as a reference for the stage crew.

The underside of the stage overhang is plaster on lath. The upstage wall of the pit does not have a permanent drape track for adjustable acoustics. The rear wall is a vertically oriented, wood slat system with gaps between the slats over black fiberglass insulation. This provides an estimated

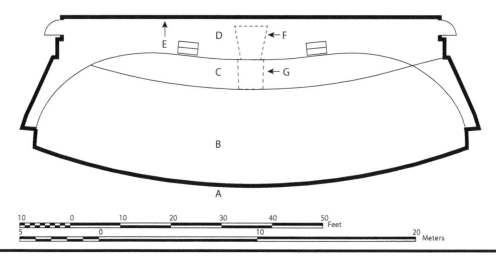

Figure 9.7 Metropolitan Opera House Pit, New York, NY, 1966. Often considered the gold standard of orchestra pit design. (A) Solid pit rail. (B) Main pit lift used for large-scale operas and ballet. T&G oak wood strip floor on oak subfloor on steel plate on steel frame. (C) Mozart lift, used for small orchestras and (B) brought to seating level. Same floor construction. (D) Lower pit level, often filled with temporary risers to raise to C level. T&G oak wood floor on oak subfloor on sleepers over concrete. (E) Upstage wall, 1/2 in. (13 mm) thick vertical wood slats, 4 in. (100 mm) wide, and a 1/2 in. (13 mm) gap between with glass fiber behind. (F) Permanent prompter's box. (G) Temporary prompter's box.

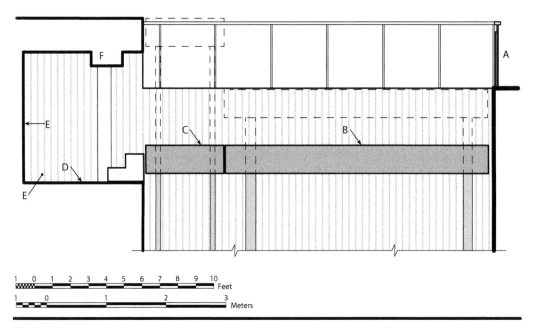

Figure 9.8 Metropolitan Opera House Pit, New York, NY, 1966. (A) Solid wood pit rail. (B) Main pit lift at typical play elevation. (C) Mozart pit lift at typical play elevation. (D) T&G oak wood floor on sub floor. (E) Wood slat rear wall.

NRC of about 0.7. When additional acoustic absorption is required by the music director or conductor, velour drapes are temporarily rigged to cover the slat wall.

Myths and Misconceptions

Myth #1: Nothing Can Be Done Acoustically If the Pit Orchestra Is Too Loud and Overpowers the Singer(s)

False. I have found that reconfiguring musicians within the pit can significantly help. For example, one can relocate the higher-energy brass and percussion to a position under the stage area, use carpet pieces under percussion, and extend acoustic drapes behind high acoustic energy instruments in order to temper sound.

Myth #2: Expensive Mechanized Pit Lifts Can Be Eliminated to Save Costs and Are Readily Replaced with Manual Pit Platforms

False. This is a very costly error in multi-use hall design. The mechanized lift saves significant operational labor costs and improves acoustics. It serves as an excellent large elevator for transporting pianos and percussion to the pit and for moving audience seating to and from storage. A mechanized lift allows for quick reconfiguration of the hall when the orchestra is to play out on lifts in symphonic mode. Manual platforms are heavy, cumbersome, and dangerous for technicians to move. I do not recommend manual pit platforms (see Figure 9.9).

Figure 9.9 Alice Tully Hall, New York, NY, renovated 2009. To foster quick changes to hall configurations, seating wagons (wheeled platforms) store under the stage. (Photographed at Lincoln Center for the Performing Arts in New York City.)

Myth #3: The Ceiling of the Hall in the Area Over the Pit Should Be Shaped to Project Sound to the Last Row of Seating

Incorrect. The pit area ceiling should be designed to modestly slope forward. It should be shaped to return 25%–40% of reflected sound back to the pit musicians to aid hearing. This allows the orchestra–singer balance to be controlled by the conductor and restrains the orchestra's loudness in the hall.

Myth #4: A Single Pit Lift Is All That Is Required. Two Lifts Are Redundant and Extravagant

False. Two or even three pit lifts can reduce labor costs and time for reconfiguration of a multi-use hall. Multiple lifts allow the pit to be easily set at a small configuration for Broadway productions or Mozart operas and enlarged for a large configuration such as Wagner and Tchaikovsky operas. Multiple lifts maximize seat count and ticket revenue. If the pit is designed for a large opera orchestra with 90 or more musicians, it will be cavernous for a Broadway production and push audience members far away from the stage.

Myth #5: A Well-Designed Opera or Ballet Pit Will Not Require Sound Amplification Systems

Partially true. Although natural acoustic pit sound should never be amplified for the audience, sometimes, the pit sound is mic'd and distributed through loudspeakers, or stage monitors, to the stage. This aids performers both on and off stage and helps the stage crew to follow the score. Care must be taken to keep amplified sound levels inaudible to the audience.

Chapter 10

Acoustic Design of the Stage

Introduction

The stage area of a multi-use hall is an acoustically critical space because it is the point from which sound emanates. The stage is the heart of the sonic performance from which the energy and vibrations travel to the audience.

The Importance of Stage Acoustics

Creating a successful stage design is more complex than one may think; there are many factors that contribute to a good stage. Sound energy should flow freely between the stage volume and the hall volume. In some cases, the sound on stage circulates above the stage ceilings and around the walls in a reverberant volume or in chambers to emerge later in time as reverberant warmth and added bass response to both stage performers and the audience.

Many of the ideas you will read about in this chapter are derived from listening to choral, chamber, and orchestral musicians. I have had and continue to have the delightful experience of listening on dozens of stages in various locations, while highly accomplished musicians ply their art. I have been able to listen to what they hear and, more importantly, what they cannot hear.

Hearing and Listening

Hearing and listening are two distinct skills. Hearing intelligently takes years of ear training. Hearing the sound vibrations created by instruments or singers when on stage only scratches the surface. It is even more critical to listen to the time arrivals of reflections from ceilings and walls and feel the vibration of sound through the floor. Hearing music is very different from listening to the hall respond after the music stops or listening to the running liveness during the music. The acoustician must listen, blend, and balance anomalies caused by the shell or stage positions of musicians.

Any acoustician that wants to know how stages work must listen carefully to a number of locations on the stage—the brass section, woodwinds, the back of the violin section, and near the conductor with the cellos. It is important to discern what those artists can hear—of other sections, of other musicians in their sections, and of the hall itself. Only through this careful and studied immersion is it possible to learn about how these spaces actually function—beyond the theory. Form your own opinions on what works and what does not.

What to Listen for on Stage

- The ability to hear instruments clearly across the stage during heavily scored passages.
- The ability to hear the second violins from the bass section, while the brass is playing.
- Does percussion drown out other nearby sections or are they clearly audible?
- Can the woodwinds hear the bass?
- How are the communications between bassoons and the bass?
- Can the French horns hear the rest of the ensemble?
- In choral works, can singers on the outer edges hear the singers on the far side away from them?
- What are choral and orchestral balances like on the stage? In the choir?

Design of Orchestra Shells

As and acoustician, I favor an orchestra shell, or a concert enclosure, that provides exceptional sound clarity and transparency over the sound of a blended or homogeneous sound. We like to clearly hear the attack of the bow on the violin string or the slap of the pad of the clarinet and the percussionist's triangle within the totality of the music. We believe that the audience member should be able to pick out the sound of an individual instrument in a thickly scored piece, as well as be able to listen to the totality of the blend and richness of the sound. Truly an individual preference, the following design narrative describes a design that provides clarity, presence, and clear articulation of instruments rather than the traditional concert hall sound where individual instruments are blended into the whole.

If the sound on stage is so clear that the performers can hear each other, the audience benefits as well. Perhaps it is the musician in me that wants to be able to hear the flute part and listen to the fine musicianship and artistry. One conductor told me that he would prefer a shell acoustic where he can control when an individual instrument is clearly audible or not and prefer the shell sound to be more blended and full. In any event, the reaction to Jaffe Holden's approach to shell design has been overwhelmingly positive from both musicians and audiences.

I subscribe to the belief that many people listen with their eyes, and if the shell or hall looks odd, then it would not receive the same level of positive response even if the sound is identical. For example, I had an architect who was struggling with a shell design aesthetic ask me if he could just paint it totally black and hang sound-transparent metal mesh in front of the shell towers. I told him in no uncertain terms that *no*, the musicians would never accept such a design even if we could make it sound fabulous. That led to a long discussion on the idea of psychoacoustics, that is, those features and aesthetics that are not actually acoustic but that affect the acoustic impression of the sound.

Lightweight Acoustic Shells

Much of what is expressed here is based on the work of my mentor, Chris Jaffe, and his guidance over 30 years. He and I had many collaborations with shell design—he started his career as a shell manufacturer in a company called Stage Craft Corporation. He made acoustic shells out of aluminum frames and fiberglass sheet for a durable, lightweight, and weather-resistant design, as many were used outdoors. Jaffe's company achieved extensive success, and I heard tales of how they were used on the White House lawn by the National Symphony Orchestra in the 1960s and were placed at the State Theater at Lincoln Center for the New York Philharmonic.

Sizing the Orchestra Shell

The gross area needed for an orchestra shell is usually calculated from the total largest ensemble that is expected on the stage. Generally, this is a full symphony orchestra with over 100 musicians plus a seated choral group. Perhaps this is reminiscent of the cliché that one should not design the church just for Easter Sunday, but the acoustician should be aware that the client will at some point fill the stage with 300–350 musicians with seated chorus.

Use the basic starting figure of 25 ft.² (2.3 m²) per musician to get a rough estimate. Do a stage layout to obtain an exact measurement (see Figure 10.1).

Note the gradual trapezoid flaring of the side walls of the layout at about 10° off the centerline and sending sound in the direction of the audience. This avoids having parallel walls that can cause acoustic anomalies and resonances. If the angle is too drastic, it will project too much sound off the stage and result in a reduction of onstage and cross-stage hearing functions.

The stage at its widest should not exceed 60 ft. (5.6 m) at the front. A wide stage will force the musicians to spread out and will hamper the ability to hear onstage. If there is a proscenium, then the shell should be about 55–57 ft. (5.1–5.5 m) wide at the proscenium opening, allowing the orchestra to grow to a full 60 ft. (5.5 m) on the stage apron. Often, there is a desire to have the orchestra out on lifts in the hall to better acoustically couple the orchestra to the room. The proscenium opening could thereby be reduced further to 53–52 ft. (4.8 m), which would benefit most nonsymphonic uses.

Rarely is the orchestra shell too deep acoustically; perhaps visually it can be overwhelming, but sonically having extra space at the rear behind the ensemble is not a problem if the shell side walls and ceilings are correctly designed. For years, we designed orchestra shells whose rear walls could be positioned further downstage in an attempt to make the shell smaller and acoustically tighter. Later, I found that orchestras preferred the empty space and noticed that the balance of high-energy instruments to strings was improved. Brass, percussion, and particularly French horns liked the added space, so I no longer advocate for moving walls downstage around smaller ensembles. Again, this only works when the side walls are narrow, and the ceilings of the orchestra shell are designed for projection, blending, and onstage hearing.

Orchestra Lifts

The ideal multipurpose hall is one where the orchestral musicians are not separated from the hall by the proscenium arch but rather inhabit the same space as the audience. Ideally, we should move the musicians forward within the hall on the stage apron and orchestra lifts (see Figures 10.2 and 10.3). While acoustically superior, this design direction gives rise to a host of sight line issues, complexity in the forestage ceiling, interference with theatrical lighting, additional lifts, seating wagons, and perhaps even additional excavation. We believe that the results are well worth the additional effort and cost over a more standard orchestra shell design, borne out of the fantastic response by both artists and audiences to our multi-use hall designs at the Schuster Center in Dayton, Ohio, the RiverCenter for the Performing Arts in Columbus, Georgia, and the Cannon Center for the Performing Arts in Memphis, Tennessee (see Figure 10.4).

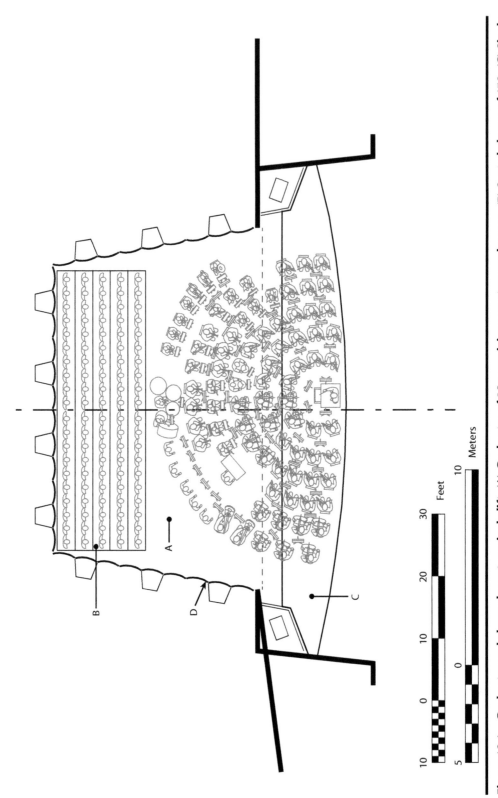

Figure 10.1 Orchestra and chorus layout on single lift. (A) Orchestra of 90+ musicians on stage and apron. (B) Seated chorus of 150. (C) Single lift brings most of string section past the proscenium arch. (D) Ten identical rolling shell towers with diffusive face as tall as proscenium opening.

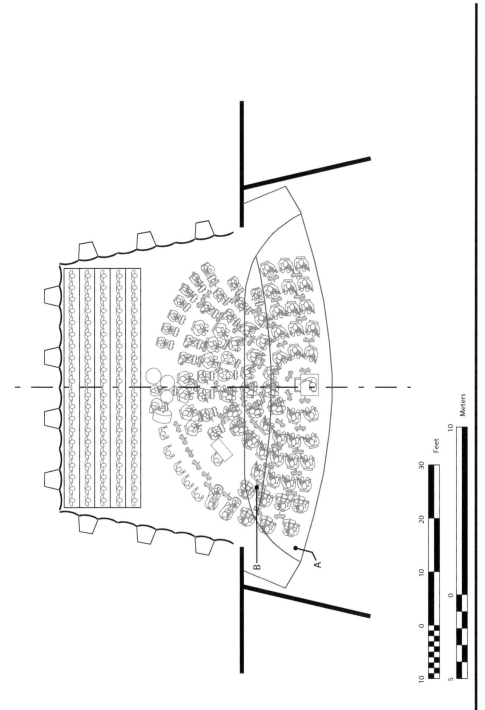

Figure 10.2   Orchestra and chorus on two lifts. (A) The second lift allows more flexibility for the size of the pit orchestra and extends the orchestra and string section further into the hall. (B) The Broadway pit lift is used for smaller amplified ensembles. Note that the forestage reflector over the orchestra extends further into the hall.

Figure 10.3 The Dayton Philharmonic at the Schuster PAC, Dayton, OH, 2003. Neal Gittleman, artistic director and conductor. Note the position of the orchestra, risers under rear violin sections, and forestage reflector overhead. The Schuster PAC is a featured case study.

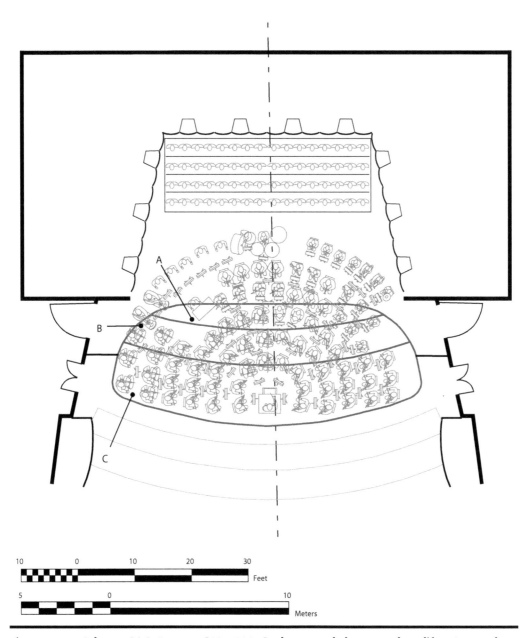

Figure 10.4 Schuster PAC, Dayton, OH, 2003. Orchestra and chorus on three lifts. (A) Broadway pit lift. (B) Second lift/pit for large opera orchestra. (C) The third lift extends 2/3 of orchestra out into the hall. Seat wagons on this lift store upstage. Note that the large movable forestage reflector is critical for this scheme.

Forestage Reflector

The large movable forestage ceiling reflector is vital to the success of this design direction (play in out on lifts) and is the most complex component. It must fly about 35–40 ft. (10.6–12.2 m) above the stage lifts for aiding onstage hearing, blending, and projection. It must also have the ability to retract up to clear theatrical catwalk lighting positions for all other productions.

At the Schuster Center, this ceiling element stores vertically in the forestage area under the grid. When needed, it slowly descends from the forestage area in a vertical position and finally tips horizontal at the appropriate location and angle (see Figures 10.4 and 10.5). See the Shuster Center case study (Appendix III) for a detailed look at the acoustics.

The forestage piece must meet the following criteria:

- Cover the entire zone where the orchestra will play, including far stage left and right.
- Be more acoustically open than a typical shell ceiling so that sound can travel through it and engage the volume above the musicians. If it is too solid, it will throttle the sound. Typically, 25%–30% should be open, and elements should not be smaller than 6 ft. (1.85 m) in any direction.
- Project sound out to the audience but at a less severe angle than the onstage reflector since the audience is closer.
- Be convex and made of honeycomb or molded materials so that it is lightweight and has appropriate stiffness. This allows low frequencies to pass through it to the upper volume of the hall.
- Be tunable in the field in both angle and height.
- Be on motorized winches with presets.
- Incorporate orchestral lighting and provide openings for main array loudspeakers to pass through it.

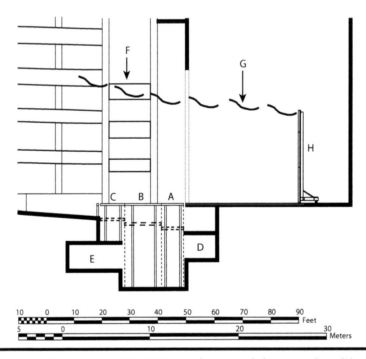

Figure 10.5   Schuster PAC, Dayton, OH, 2003. Orchestra and chorus on three lifts. (A) Broadway pit. Seat wagons here store under the pit in storage area, (D). (B) Lift 2. Seat wagons on this lift store in area E. (C) The third lift goes from seating level to stage level only and wagons store upstage. (F) Three forestage reflectors on rigging store above the ceiling. (G) Shell ceilings on stage store in fly loft. (H) Rolling shell towers store off stage in alcove.

The stage extensions are a series of two or three lifts to extend the stage 30+ ft. (10.6 m) into the hall. They can include seating wagons to allow a quick change from stage configuration to audience seating and storage garages below to store the wagons. Theatre Projects Consultants engineered the movable systems that make all these complex systems function at the Schuster Center and the RiverCenter and provided invaluable collaboration.

Note that the quantity of shell walls and ceiling reflectors required onstage with this orchestra lift design is much reduced, as is the shell storage area. The onstage units are typical orchestra shell units, as described in the section below.

Reviews for this design have been ecstatic and well worth the added complexity and cost. Janelle Gelfand (2003) of *The Cincinnati Enquirer* wrote, "By the conclusion of Thursday's concert of Beethoven, Mendelssohn, Vaughan Williams and Stravinsky led by music director Neal Gittleman, it was clear; this (the new Benjamin and Marian Schuster Performing Arts Center) is one of the most stunning acoustical spaces in Ohio—perhaps in the nation."

Orchestra Shell Walls

The question of the material and weight of shell walls is one that is open to much discussion, as is the shape and the height of the shell. Real-world design results drive decisions, not theories of what *should* work. The onstage acoustic performance of halls with shell designs around the world reveals how best to design a cost-effective and functional orchestra shell.

Theater and stage consultants have much to add to this discussion. Shell towers or walls must be movable and storable in a multi-use hall to clear the stage for other uses. This might be accomplished by rotating the side walls open to allow access, as was done at Alice Tully Hall. Please read the Alice Tully Hall case study (Appendix I) for more information. Walls can also be flown out on rigging up into the stage house or tracked offstage on overhead systems similar to those that move air walls in convention centers or ballrooms.

For most halls, I advocate putting the side and rear walls on large dollies with castors that allow units to roll on and offstage. These systems are built of steel or aluminum frames with heavy counterweights on the base to balance the massive height. Towers standing 30–35 ft. tall (2.7–3.2 m) are not unusual in multi-use halls, but they can be moved quickly by two or three people to a storage area offstage. Towers can nest horizontally within one another and may have one or two side wings to be larger in plan without adding more bases. This allows efficient storage offstage in as small a footprint as possible (see Figures 10.6 and 10.7).

To provide added stiffness, diffuse the sound field, and promote even sound distribution on stage, the preferred shape in plan is often the large radius curved panel. It measures 10–12 ft. (3.1–3.6 m) across, similar to the Wenger Diva Shell. Many other shapes have been successful—some segmented, some asymmetrical—but this shape and unit size have good results.

Stiffness of the tower faces is an important factor to consider, especially since the mass and weight of the tower are a big concern to theater consultants and structural engineers. Many a stage floor has been crushed, scarred, or splintered by the excessive weight of a tower on wheels. We prefer the honeycomb shell structure for its stiffness and rigidity with the panel surface weight as high as the theater consultant will allow on the floor, often ¼–½ in. (6.3–12.7 mm) thick wood or laminate. The weight is low, about 2–5 lbs./ft.^{2} (9.8–24.4 kg/m^2) compared to the wall surface weight in the hall, so the acoustician must carefully consider the low-frequency response in the hall. The towers will become less-effective sound reflectors at 125-Hz octave and below.

Figure 10.6   Orchestra Shell Towers. Note the convex shape for sound dispersion and diffusion, wood veneer for an enhanced visual appearance, and integral doors for access.

With this type of shell, the boundary surfaces must be substantial with minimal low-frequency absorption to balance the fact that the shell's low-frequency response is less than the mid and highs. My experience is that the bass response in the hall is excellent when the shell absorption is considered holistically with the hall design. This holds water even if the density of the shell towers and ceilings is substantially less than the hall.

Recall that we are dealing with real-world and evidence-based acoustic design here. Massive-weight shells have proven to be expensive, unwieldy, and costly to set. They also tend to destroy the floor and are disliked by stage crews and hall management. What counts is the overall acoustic performance, balance of instruments, and quality of the sound in the shell and hall system. What is the difference if the G is 2–3 dB less in the hall due to shell absorption if the percussion-to-brass balance to strings is excellent? We believe that venting high sound energy to the stage house actually improves the hall's acoustics by reducing harshness with loud passages. We follow the same criteria with the ceilings of the shell, as discussed in the section Orchestra Shell Ceilings.

Important Considerations with Shell Towers

- *Wood surfaces, including thin veneers, are preferred by musicians and management for the shell skin over painted surface materials such as fiberglass or metals.* This may seem illogical, but recall that musicians are artists at work, and their perception of the environment affects the results. Chris Jaffe used to say, "Wood is good and sound is round."
- *Avoid significant areas of high-frequency diffusion materials.* Specular reflections aid onstage hearing, and excessive diffusion works against them.

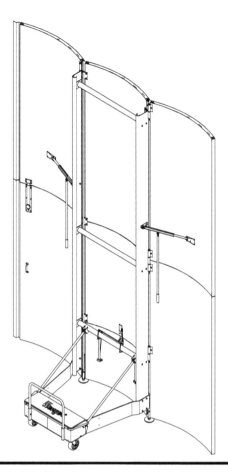

Figure 10.7 Isometric view of shell tower. The center counterweighted base supports two folding wings. Note the convex curve for diffusion, entry doors, and ability to nest units for storage. (Courtesy of Wenger Corporation.)

- *There is always a gap at the fire curtain in multi-use halls; do not be alarmed.* Not every ounce of sound energy is required.
- *All towers should be the same height rather than tapering toward the upstage for interchangeability and flexibility.* Stages deflect under the tower load, so it is critical to level feet on towers. Shell manufacturers provide excellent technical design support for standard and custom tower design. I have used Wenger, Stage Right, and J.R. Clancy frequently.
- *Provide as many double-door entrances as possible for ease of stage equipment access.* After all, a concert grand piano needs a 6-ft. (1.8-m) clear opening.

Orchestra Shell Ceilings

Designing the ceiling of the shell is incredibly complex and can be an overwhelming task for a novice. I panicked for months early on in my career when I was assigned the task of designing an orchestra shell ceiling for a community college. Thankfully, I had an excellent mentor and evidence-based experience to draw upon.

Functionality

A shell ceiling system provides a blending chamber that yields a balanced cohesive sound. It serves as a tunable projection device that sends clarity and articulation information about early reflections to the audience. The system helps musicians hear themselves onstage and balances high-energy instruments by bleeding off sound. Additionally, the system allows for air-conditioning and cooling to exist beneath hot lights and houses the lights needed for reading sheet music and illuminating artists.

The ceiling can be raised or lowered, and the angle can be adjusted toward or away from the audience. Tipping the ceiling sends more sound out to the audience, but at the same time, it often reduces the levels of the musicians below and affects onstage blending and balance. Lowering the ceiling usually aids onstage hearing, but it affects balances by overemphasizing sound from musicians. It may also look out of place from the audience's perspective.

Materials

Use honeycomb or a similarly stiff double panel with a weight of 2–5 lbs./ft.^{2} (9.8–24.4 kg/m^2). The material must be easy to adjust. Remember, wood is good. Veneers or solid wood works well.

Size

The dimensions should not be less than 6 ft. (1.8 m) to allow full-frequency reflections. Somewhere between 7 and 9 ft. (2.1–2.7 m) is preferred. Remember to provide openings for air and light either upstage/downstage or cross stage. If the ceilings are chopped up, they will leak too much sound into the stage house.

If the ceilings are over an open stage without a contained stage house, the criteria are different. In that case, the acoustician wants sound to move past the ceilings to energize the volume above. I find that openings of about 18 in. (45.7 mm) +/– 6 in. (15.2 cm) work best in terms of airflow and acoustics.

Shaping

Shell ceilings start at about 30 ft. (9.14 m) above the stage deck at the proscenium and gradually taper back to about 25 ft. (7.6 m) at the upstage. The upstage-to-downstage shape dimension is the most important. Shaping left to right is less acoustically critical. This type of shaping may be used to improve aesthetics but tends to make the shell harder to store in the fly loft.

When the ceiling is designed in a convex shape, it projects some sound out to the hall, but less than half of that sound goes back to the musicians on the stage. I have used a swoop or airfoil shape for many designs, and it has proven extremely successful for both onstage hearing and projection (see Figure 10.8).

Custom Designs

The Wenger-style shell is not for every project—especially if the architect desires a strong visual connection between the hall's design and the shell on stage (see Figure 10.9).

Figure 10.8 **Stage hand measures angle of honeycomb acoustic shell ceiling with aluminum frame during acoustic tuning. The unit angle can be adjusted to pretuned play position and can store vertically. Note integral theatrical lighting instruments.**

Shell at the Globe-News Center

The shell visually completes the hall within the Globe-News Center in Amarillo, Texas. Walls and ceiling elements mirror the hall's enveloping design in a common architectural language of shaped wood. The portable shell adheres to all acoustic design principles. The surfaces provide blending, projection, and onstage hearing reflections, and openings to the rear of the ceiling vent the higher-energy instruments and allow for air circulation and lighting positions. However, this particular shell does not allow for tuning.

The unitized wall/ceiling piece moves as a complete assembly into a storage garage upstage. A crane–rail system suspends the shell walls and ceilings off the stage floor and moves them upstage in a matter of minutes. This allows the change from theatrical mode to concert hall mode to happen very quickly in support of the Amarillo Symphony. In terms of cost, the large storage garage needed for this shell is at a premium. The cost of the shell itself is in line with other shell designs (see Figures 10.10 and 10.11).

Figure 10.9 Globe-News Center, Amarillo, TX, 2006. Unitized shell. The shell continues with the design of the hall using similar diffusive wood elements. Note orchestra positioned fully within the shell during acoustic tuning; normal play position is out on apron and lift.

Stage Floor Construction

The stage floor is a surface that has many purposes in a multi-use hall. In a pure concert hall, the floor would be optimized for orchestra, chorus, and recital use, but in a multi-use hall, the floor must work for opera, theater, and dance performances plus corporate events.

Materials

In my experience, soft woods such as pine are not appropriate for multi-use halls because of durability issues. Suitable materials include maple, oak, or ash hardwood strip flooring with a gloss or semigloss finish. A plywood subfloor made of wood sleepers such as 2 × 10 ft. (0.6 × 3.4 m) on edge, 12 in. (0.3 m) on center should be added for strength. Ensure that the floor is mounted directly on the structure below to give the space low-frequency energy from percussion and bass.

Often, the theater consultant will prefer that the multi-use hall gets a black-painted floor with a hard board surface such as masonite or plastic because it is durable and can be painted,

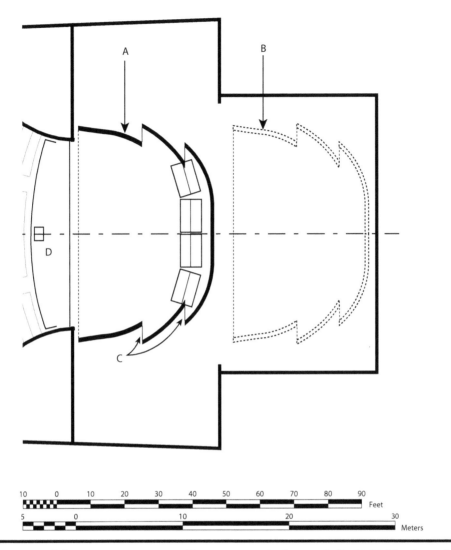

Figure 10.10 Globe-News Center, Amarillo, TX, 2006. Orchestra shell. (A) Unitized wood shell hung on rails does not touch floor. (B) Shell in storage garage. (C) Openings to vent sound also allow air and lighting. (D) Single lift/stage extension.

scuffed, hammered into, and otherwise abused. They may desire an engineered floor system from Robbins or another manufacturer that is derived from the sports floor environment (see Figure 10.12).

When symphony use is high, the floor can have a dark-stained hardwood finish as a compromise for the musical community. It could have a resilient channel system underneath; however, this is not ideal because it dampens the floor resonances. This effect can be mitigated by placing an orchestra riser system under the cello and bass instruments.

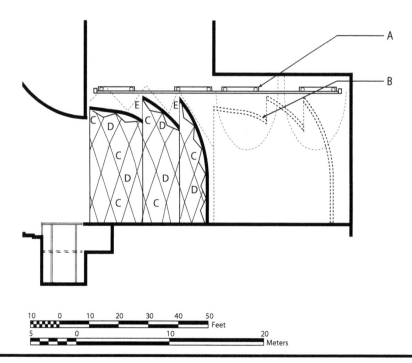

Figure 10.11 Globe-News Center, Amarillo, TX, 2006. Unitized shell. (A) Crane rails used to support shell and move into storage (B). (C) Shaped wood diffusive elements between large-scale diffusive ribs (D). (E) Openings in shell vent sound, allow air movement and lighting.

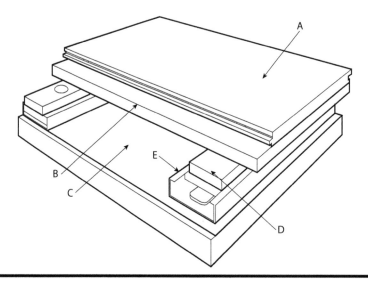

Figure 10.12 Resilient stage floor construction. Typical multi-use stage floor provides a hardwood finish that supports heavy rolling shell towers, provides resiliency for dancers and resonance for instruments. At times it is covered with a thin hard-board than can be replaced when damaged. (A) Northern hardwood maple strip flooring. (B) Structural rated sheathing subfloor. (C) Concrete structure with vapor barrier on top. We don't recommend filling the void with fibrous materials. (D and E) Sleeper and rubber pad in factory-assembled steel-enclosed system that limits overcompression.

Air-Conditioning on Stage

Perforations in the stage floor can deliver fresh air to musicians and are especially helpful in out-door summer environments. Refer to the Baldwin Auditorium case study (Appendix II) for more details on this type of system. The partially open ceiling system requires no modifications to the shell because the typical stage air supply systems heat and cool the stage by delivering air down through the openings.

Orchestra Riser Systems

There is no standard orchestra riser system. Each one is custom designed for application within individual halls. Systems can range from an expensive and custom-designed wood unit to a set of basic rectangular platforms that suit a variety of needs.

Acousticians must produce shell designs with ceiling designs that enhance onstage hearing and reduce the acoustic need for a riser system. An exception to this includes the occasional cello risers or bass risers because they increase bass sound levels. Many orchestras have good sound without riser systems at all especially when the hall design thrusts musicians out in the house.

Example: The Benjamin and Marian Schuster Performing Arts Center

The Benjamin and Marian Schuster Performing Arts Center in Dayton, Ohio, does not use riser systems. The conductor of the Dayton Philharmonic Orchestra, Neal Gittleman, described to me how he tested dozens of riser configurations for musicians and, after years of experimentation, decided upon our original recommendation—orchestra flat on the floor.

Risers at the rear of the orchestra elevate sections primarily for visual and not for acoustic reasons. If you follow the logic that people listen with their eyes as well as their ears, then making the orchestra more visually exciting makes sense regardless of acoustic results. Putting the brass and percussion on risers in the rear of a shell only exacerbates the balance issue between high- and low-power instruments. Raising these louder musicians above the others reduces the sound absorption by the musicians in the front (see Figure 10.13).

Myths and Misconceptions

Myth #1: Orchestra Shells Must Be Heavy and Massive in Order to Function Correctly

Jaffe Holden has spent 40 years showing that stiffer and lighter-weight materials like honeycomb with a hardwood face can produce excellent results when used appropriately. Understand that some sound leakage is normal and can actually be advantageous.

Myth #2: Shells Should Be Sealed Chambers with No Holes or Gaps That Let Sound Escape

Venting the shell for superior balance between sections means that there will be gaps. These gaps are useful for air ventilation and lighting. Containing every decibel of sound for maximum G can lead to harshness and distortion.

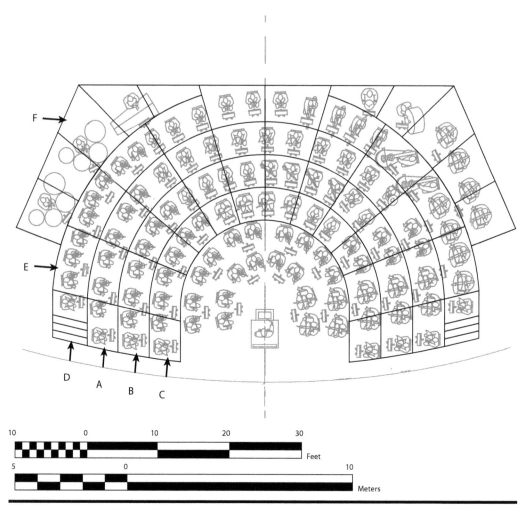

Figure 10.13 Portable orchestra riser system for multi-use hall. (A) Tall unit allows B and C levels to be stored within and rolled off stage. (D) Stair unit. (E) Taller rear riser. (F) Highest percussion level risers. Note the portable lift needed to bring pianos to upper levels.

Myth #3: Orchestras Do Not Need Ceiling Reflectors; the Great Concert Halls Do Not Have Them

In order to produce excellent performances, multi-use halls need reflectors for projection, blending, and excellent onstage hearing. Surveys prove that musicians greatly prefer halls with overhead reflectors.

Myth #4: Orchestras Should Stay Behind the Proscenium and within the Shell in Order to Achieve the Best Sound and Support for Musicians Onstage

Not true. With the appropriate forestage walls, shaping, ceiling reflections, and multiple pit lifts, a forward location can result in great sound for both performers and audiences.

Myth #5: In Terms of Shells, Wood Walls Are Acoustically Superior to Other Materials

This is not true. Wood used on shells is usually thinner than 1/8 in. (3 mm) glued to backup materials such as particle board on honeycomb or foam core. Wood this thin has no acoustic benefit; however, it is psychologically effective.

Chapter 11

Orchestra Seating, Balconies, Boxes, and Parterres

Introduction

Shaping the interior of the hall with boxes, balconies, parterres, and galleries is exciting. This is when the character of the interior gels into concrete architectural forms via the intense collaboration between the acoustician, the theater consultant, and the architect. Out of this effort comes the fundamental proportions that will define both the visual feel of the hall to patrons and performers and the acoustic response. But where does the acoustician begin this effort?

Volume Proportions and Dimensions

While the overall acoustic volume for the hall has been defined by the program, the proportions and location of this volume are obviously defined by the size and location of the seating areas plus the walls, ceiling, and floors that create the enclosure. The program may set forth the audience area in square feet or square meters. Usually, this is about 10 ft.^{2} (0.9 m^2) per person.

Realistically, the entire audience cannot be on one giant level. The audience is distributed around the room in a pleasing visually manner that allows for excellent visual and aural sight lines to the stage. There are no magic dimensions that produce fantastic acoustics for large multi-use halls with 2000–2400 seats, but there are important maximum acoustic dimensions to consider (see Figures 11.1 and 11.2).

Important Dimensions for Large Halls

- Last seat on the orchestra floor no more than 100 ft. (30 m) from stage edge.
- Last seat in the balcony no more than 135 ft. (41 m) from stage edge.
- Width of hall at the proscenium zone is not to exceed 56 ft. (17 m).
- Width of hall at throat area is not to exceed 75 ft. (23 m).
- Width of hall in rear seating area is not to exceed 105 ft. (32 m).

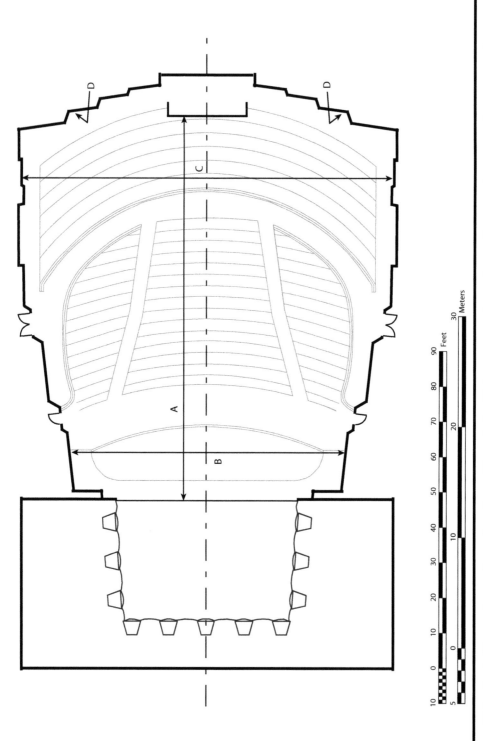

Figure 11.1 Desirable maximum dimensions for multi-use halls. (A) Last seat on the orchestra floor 100 ft. (30.5 m) from stage edge. (B) Last seat in the balcony 135 ft. (39.5 m) from stage edge. (C) Width of hall at the proscenium zone not to exceed 56 ft. (17 m). (D) Width of hall at throat area not to exceed 75 ft. (22.8 m). (E) Width of hall in rear seating area not to exceed 105 ft. (32 m).

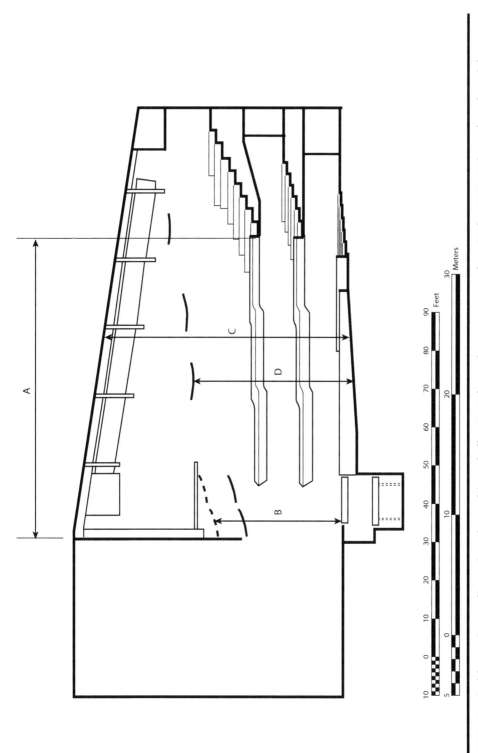

Figure 11.2 Desirable maximum dimensions for multi-use halls. (A) Balcony fronts no more than 60 ft. (18.5 m) from the edge of the stage. (B) Ceiling height over pit lift no more than 45 ft. (13.75 m). (C) Ceiling height in hall, defined by volume requirements.

- Balcony fronts are no more than 60 ft. (18 m) from the edge of the stage.
- Ceiling height over pit lift is no more than 45 ft. (14 m).
- Ceiling height in the hall is defined by volume requirements.

Lessons Learned

The width at the rear of the hall is less acoustically important than the width of the hall at the throat zone and at the proscenium.

The Wagner Noël is over 103 ft. (31 m) wide at the widest point far back in the hall. This large width was once perceived as negative by acousticians who favored narrow shoebox halls such as the Boston Symphony Hall. However, visual intimacy and loudness were improved in the Long Center by adding throat walls and bringing the ceilings closer to the stage. Widening the seating area at the rear of the hall allowed the theater consultant to bring the audience and balconies closer to the stage without losing acoustic quality and visual intimacy.

Further details are included in the case studies within this book. In Chapter 13, ceiling design is covered in more detail.

Hall Configuration

It is critical to have a discussion with the architects, theater consultants, users, and stakeholders regarding the size of the orchestra seating level and the number and dimensions of balconies and side boxes or side galleries. In the early design phase, the discussion focuses on the layout of the audience on the orchestra level, balconies, and side boxes. When discussing the hall's maximum capacity, ask the question if the maximum number of seats required includes the seats on the orchestra pit lifts or not. We have had clients demand 2000 seats in the hall, and then we find out that they meant 2000 seats in a Broadway configuration with the orchestra pit full of musicians. The actual seating capacity is then 2000 + 30–40 seats on the lift. As with most acoustic issues, when one parameter is modified, others are affected in a complex ripple effect.

Orchestra Level

Operators of these facilities often desire that the orchestra-level seating area have as many seats as possible because they produce the highest revenue, and audiences want to be on the lowest level. They perceive that the acoustics and site lines are best on this level. This, however, is false. For seats too close to the stage, the direct sound overwhelms the reverberation, spaciousness, and lateral reflections. Seats in the rear of the orchestra suffer from the loss in G levels as sound passes over the sound-absorptive seating zone at a low angle. The orchestra seating is farthest from sound-reflecting walls, soffits, and the ceiling and may result in low levels of C80, lateral reflections, and intimacy. Design the hall's walls and ceiling to upgrade the acoustics of the orchestra seating so that it does become acoustically excellent and worth the high price they command. In fact, the best seats, acoustically, are usually the first two or three rows in the balconies.

Limiting the orchestra level seating to 800 is a good rule of thumb. When reaching 1000 seats, the hall will often be excessively wide or deep. We advocate adding a parterre level that can break up the large mass of seats and make the seating plan seem smaller and more acoustically intimate but usually at some loss of seating efficiency and maximum capacity. Advocate for seating on the orchestra level to include two or more interior aisles instead of continuous rows, formerly called

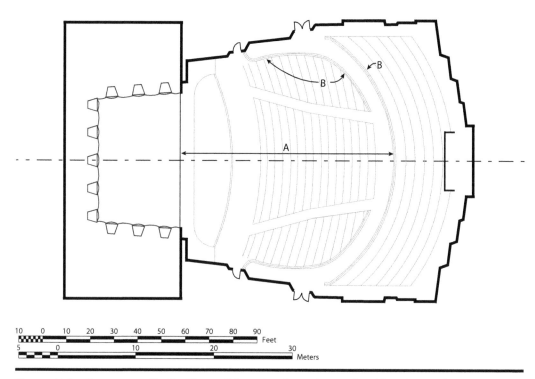

Figure 11.3   Parterre wall. (A) The wall location in orchestra level breaks up the seating area into smaller segments, raises rear seating areas for better line-of-sight to musicians, and provides a modest reflective surface within the seating zone. (B) Two parterre walls at Wagner Noël.

continental seating. This has less to do with the acoustics of the hall and more with providing audiences the ability to freely circulate and socialize before the show begins and at intermission (see Figure 11.3).

Parterre

Raising the rear and sides of the orchestra seating area into a separate seating zone is useful in creating a more intimate hall because it reduces the vast sea of long, uninterrupted rows. Long rows tend to emphasize and exaggerate a hall's width, and the parterre level can visually fix this issue.

The acoustic value of the parterre and the parterre wall is minimal to the seating on the main orchestra floor. The surfaces are small in relation to the side walls, which limits the energy and frequency response, and the sound that actually impinges on the wall is slight if an audience is seated in front of it. It can be a negative surface if it is a smooth sound-focusing surface that returns sound to the stage as an echo. Note the convex shape and diffusion applied to the parterre wall at the Wagner Noël Hall.

The real value to a parterre is that the seats in the first few rows are raised above the plane of the audience's acoustic absorption and, therefore, receive more direct sound. A very steep parterre can work against good acoustics if it is so steep that it limits the sound returning off the rear wall of the hall. The sound returning off the lower rear wall is critical to the musicians' feedback sense of envelopment for the audience in the orchestra seats (see Figures 11.3 through 11.5).

Figure 11.4 Schuster PAC, Dayton, OH, 2003. Plaster parterre wall is both angled back and diffused into small segments.

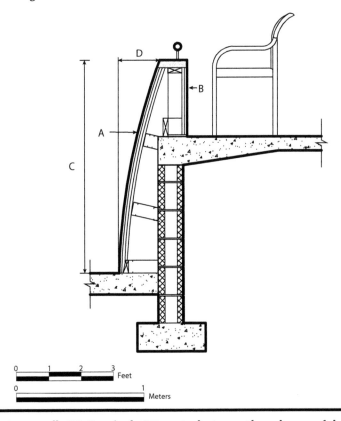

Figure 11.5 Parterre wall. (A) One inch (25 mm) plaster or three layers of drywall shaped in convex manner. (B) Rear of parterre can be plaster or two layers of drywall. (C) Wall can provide useful lateral reflections if over 4 ft. (1.2 m). (D) Dimension is 8–16 in. (0.2–0.4 m).

One Balcony

The one-balcony, cinematic-style hall has long been abandoned as acoustically successful for a multi-use hall. Many seats were buried deep under the balcony in an acoustic volume, which was separate from the volume above the balcony. Evenness of sound, surround, envelopment, and spaciousness for these underbalcony seats was not possible in this configuration. It should be noted that historic movie palaces with large single balconies *can* be successfully converted into a multi-use hall using electroacoustic solutions. This is highlighted in the Richmond CenterStage case study (Appendix IX) and discussed in Chapter 16.

A single balcony can be a very successful and cost-effective model for acoustically excellent halls with less than about 1200 seats. The area below the balcony must be limited to four or five rows, while six to eight rows can be well distributed in the balcony. As the seat count rises above 1200, and the hall width remains modest, more seats will be positioned under the balcony and up in the rear of the balcony. As we reach 1500 seats, we approach the limit of a single-balcony intimate hall and must advance to the two-balcony configuration (see Figures 11.6 and 11.7).

The Dallas City Performance Hall with 750 seats is another example of a single-balcony flexible hall that works exceedingly well acoustically and is described more fully in the case studies (Appendix IV). Scott Cantrell, music critic for the *Dallas Morning News* (2013), told me that he "loves this hall." When halls are a modest size, under about 1200 seats, a single balcony is the most effective way to deliver excellent acoustics for all programs.

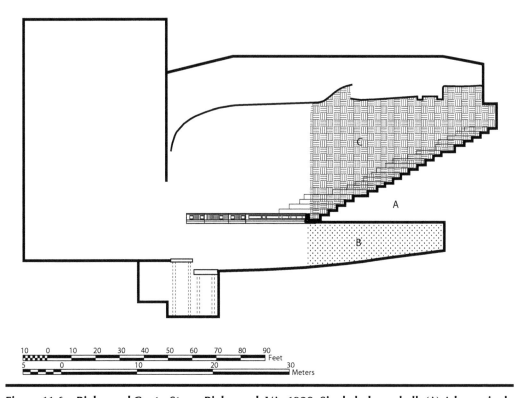

Figure 11.6   Richmond CenterStage, Richmond, VA, 1928. Single balcony hall. (A) A large single balcony creates problems by dividing the volume into two distinct acoustic spaces, (B) and (C). Energy flow between them is blocked creating nonuniform RT and reflections.

Figure 11.7   Dallas City Performance Hall, Dallas, TX, 2012. An excellent example of a single balcony, 750-seat hall. Note that the balconies float off the walls, allowing sound to travel behind, thereby enhancing the acoustics on the orchestra floor.

Two-Balcony Halls

In order to bring the audience closer to the stage for increased G levels plus acoustic and visual intimacy, the vast majority of multi-use halls in the 1600–2400 seat range have two balconies. Three-balcony halls are less common and they may have higher construction costs than two-balcony rooms for the same seat count. This is due to the increased stairs, circulation, and elevator stops. However, we have designed acoustically outstanding three-balcony multi-use halls in Dayton, Ohio, at the Schuster Center and a hall in Lubbock, Texas, due to open in 2019. Bass Hall in Fort Worth also has three levels, but a box tier comprises the first level there with two additional balconies. The steepness of the top level can become excessive, especially for older patrons, but symphonic acoustics in the upper levels of a hall are usually excellent owing to higher levels of reverberation to direct sound (see Figures 11.8 and 11.9).

Ratios

The old ratios of 1:1 or 1:1.5 for balcony depth to balcony opening are outdated and too stringent. These ratios create halls with two or three rows under the balconies that are not intimate and tend to push the audience farther from the stage than is necessary. For example, I believe that a seating row on the main floor located five rows under the balcony is sonically superior in achieving the acoustic criteria goals than a seating row in the back of the upper balcony.

Figure 11.8 Wagner Noël PAC, Midland, TX, 2009. A two-balcony configuration for a 1800-seat hall is an economical design. Side seating galleries provide visual intimacy.

Figure 11.9 Schuster PAC, Dayton, OH, 2003. A three-balcony design brings seating closer to the stage, increased visual intimacy, and excellent acoustics for all programs.

The theater consultant and the acoustician must be cognizant of the upward sightlines to the top of the proscenium arch from the last seating row. This will, in turn, allow sufficient acoustic spacing for high-quality C80 reflections and G levels.

Four on the Floor

In my experience, no more than four or five rows should be under the balcony on the orchestra floor (and perhaps on the parterre level), as defined on the center line. One more row might be added in the back corners of the orchestra level if those patrons can visually and acoustically see the bottom of the proscenium arch. This four-to-five row limit is important not just for the acoustics of the seats in this location but also because we believe that the reflections back toward the stage from the wall behind those seats are critical to the acoustic envelopment for all the seats on this level. Additionally, diffused and blended reflections must reflect off this wall and travel back to the stage within about 200 milliseconds after the direct sound to give the musicians a strong sense of the sound of the room. The rear wall, which will be discussed in Chapter 12, must not be absorptive, or returning reflections will be too low in sound pressure level to be perceived.

Seven in Heaven

A maximum seven row deep first balcony, also called a mezzanine, can work well if the last row of seats can view the top of the proscenium. Perhaps the seven-row maximum criteria seem arbitrary, but our experience shows that in a multi-use hall, eight rows are too deep. This is because in order for the eighth row to have proper connection to the hall and for it to see the top of the proscenium, the second balcony would need to be pushed higher and, therefore, steeper, resulting in poor access and sightlines. Seven rows or less are a good balance. It works well and is less stringent than the old established ratios of 1:1.5 (see Figure 11.10).

Upper Balcony

Acoustically speaking, the upper balcony should be as close to the stage as possible, be given reasonable sightlines, and have modest steepness as defined by the theater consultant. Here advocate for long rows similar to the old continental seating style, accessed from the ends rather than the steep aisles in the middle of the seating bank. Steep aisles require rails that block sight lines and sound lines. The ceiling over the upper balcony should send some reflections to these distant seats to increase G and C80 levels either with acoustic cloud reflectors or a closed ceiling design.

Sound-Transparent Balcony Allows Flexibility and Options

A sound-transparent balcony can be utilized to increase the permissible number of rows under the balcony beyond the four-on-the-floor and seven-in-heaven rules. By acoustically opening the balcony with a series of vertical grilles at the seating risers and using a sound-transparent ceiling below, the four on the floor row limit can be extended to six or seven rows, and the seven-in-heaven limit increased to 10 or 11 rows. This is especially important when hall capacity reaches 2300–2400 seats, and there is a desire for acoustic excellence for symphonic and operatic performances (see Figure 11.11).

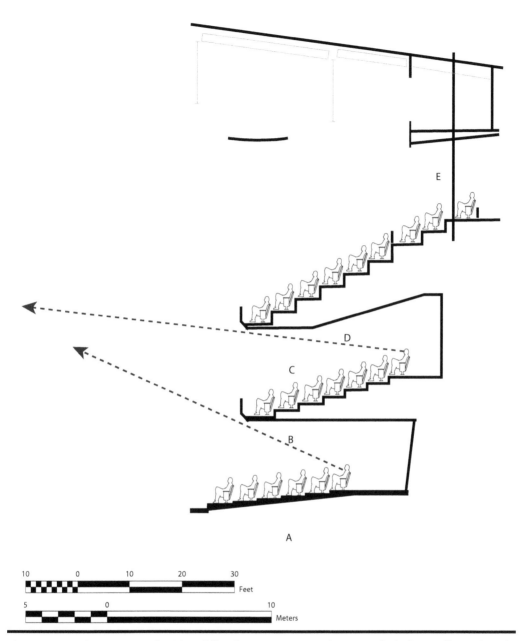

Figure 11.10 Wagner Noël PAC, Midland, TX, 2009. Maximum underbalcony dimensions in large multi-use halls. (A) No more than four rows under the balcony on the main floor. **(B)** Last row sightlines to proscenium opening. **(C)** Seven rows maximum on first balcony (mezzanine). **(D)** Last row sightlines to top of proscenium. **(E)** Three rows maximum under soffit or spot booth in balcony.

Figure 11.11 **Dell Hall at Long CPA, Austin, TX, 2008. Sound transparent balcony. (A) Seating underbalcony on orchestra level can increase from 6 to 7 rows with sound transparent balcony; (D) Seating rows can increase to 11 rows. (B) Open grilles in vertical riser. (C) Ceiling grilles allow sound to pass through to seats below.**

Side Boxes and Side Galleries

Since the 1990s, theater architects and consultants have advocated for side boxes in halls. Side boxes create additional seating, add visual interest, and bring the audience in greater contact with the performer(s) and the rest of the audience. Some claim that halls are significantly disadvantaged by the lack of boxes or side galleries. However, there are many examples of acoustically fine halls without side seating areas. There are also legitimate concerns about the audience's ability to hear and see the entire stage from side seating in a proscenium configuration.

Soffits

The soffits under the side seating area can be useful for returning lateral reflections and aiding C80 to the orchestral seating but only if the hall is not too wide, and the reflections return to the center of the hall as opposed to the rear.

The soffit and the wall are doing the acoustic work, so they need to be large enough to also serve as a full-frequency and specular reflector. A highly diffusive wall next to the soffit will not produce useful reflections (see Figures 11.12 and 11.13).

Boxes that come all the way forward to the stage may block loudspeakers and video screens. These should be pulled back from the proscenium.

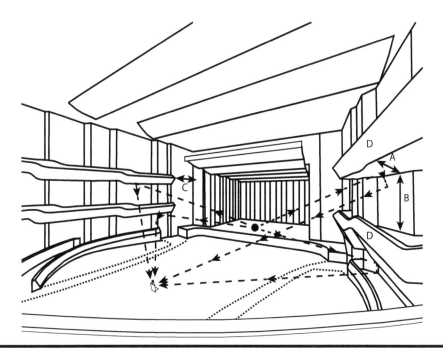

Figure 11.12 Dell Hall at Long CPA, Austin, TX, 2008. Soffit reflections. Reflections from the stage and pit reflect from soffit and wall to return to central seating area. (A) Soffit area must be large enough to be effective, minimum 6 ft. (1.8 m). Too deep, however, and sound will be trapped. (B) Height of floor to soffit must not block reflections. (C) Zone between boxes and proscenium to be clear for loudspeakers, screens, theatrical lighting, and other technology. (D) Balcony fronts are too small to provide positive reflections, but should be diffused to avoid echoes.

Figure 11.13   Bass Performance Hall, Forth Worth, TX, 1998. Soffits and sidewalls under the box tiers and balconies provide vital lateral reflections. While the walls are diffused, soffits intersect the walls at 90° with minimally applied diffusion.

Box and Balcony Front Shaping

The ideal shape of the front of the balconies, mezzanines, and side boxes has been debated and discussed for years. My opinion is that this surface is really too small in the vertical dimension, usually under 3 ft. (0.8 m), to produce acoustic reflections that increase G or aid early reflections. A gentle convex or bullnose shape that eliminates focusing echoes back to the stage from amplified sources is recommended. More complex geometries that incorporate architectural lighting, theater lighting, and projector mounts are possible and workable. Acoustically, we simply require that it be

Figure 11.14 Baldwin Auditorium at Duke University, Durham, NC, 1930. 2013 Auditorium balcony fascia. Wood surface is shaped to be diffusive to reduce echoes back to the stage, rather than direct reflections into the audience. Integral lighting for visual interest and additional high-frequency diffusion.

solid and massive (thick wood, plaster, precast Glass Fiber Reinforced Gypsum, or drywall) and have a convex or diffusively shaped surface (see Figure 11.14).

Myths and Misconceptions

Myth #1: The Side Boxes Must Extend All the Way to the Proscenium Arch to Provide Excellent Acoustics

False. The boxes should, in fact, be held back from the proscenium in order to allow important space for the main array side speakers or line arrays and video displays. Also, the acoustic and visual site lines are poor for these seats so close to the proscenium, and when the boxes are held back 15–20 ft. (4.5–6 m), the C80 acoustic soffit reflections will not suffer.

Myth #2: When the Rows under the Balcony Are Limited to Three or Four, the Acoustics in a Multi-Use Hall Are the Best

While it is true that the fewer seats under the balcony, the better, when seat counts are large (over 2000), we prefer four on the floor and seven in heaven rather than a very large upper balcony where the audience is distant, and there are only three rows in each balcony.

Myth #3: Never Have More than Seven Rows Under a Balcony

False. It is possible to have additional rows under the balcony by the careful implementation of the sound-transparent balcony design, as was done at Dell Hall.

Myth #4: Boxes with Loose Seats Are Acoustically Superior to Shallow Balconies with Fixed Seats

Well, not really. Big individual and corporate donors demand the exclusive loose seats in the boxes. These seats are less efficient in floor area. Therefore, to achieve a fixed seat count, the hall must have more rows, usually acoustically poor rows added to the back of the balconies, or it must grow in width beyond the recommended dimensions.

Chapter 12

Wall Shaping

Introduction

My business associate and friend Sig Hauck often tells me, "The devil is in the details," especially when we are negotiating business deals and contracts. This devil is also in the architectural details of walls and ceilings. If these details are wrong, the acoustic design will be compromised.

How Wide Should Halls Be?

There has been debate among acoustic practitioners about the ideal width of concert halls. Some believe in not exceeding a width of 75 ft. (22.8 m), whereas others favor far wider halls. Great halls with a narrow width do exist, including the Musikverein in Vienna, and great halls with a wide width also exist, such as the Concertgebouw in Amsterdam.

Small to Medium Halls

It is ideal for smaller halls between 500 and 1000 seats to keep the entire hall width at about 70–80 ft. (21.3–24.3 m). This allows for side reflections of laterals to increase C80. In large halls, the dimensions of the throat area, or the first third of the room, should be kept in this range. This width should be maintained at the lowest levels of the hall up to the second balcony or to the height of the proscenium. Above this height, the hall can have more width without negatively affecting C80 and loudness. Halls can expand 90–100 ft. (27.4–30.5 m) in the upper areas to add volume for reverberation time (RT). A lower ceiling and roof height for the same volume also reduces construction cost.

Larger Halls

As a hall increases in capacity to 2000 or more occupants, the dimensions obviously grow. The desire to keep the audience close to the performer drives width to be ever wider. The width can expand to 90 or even 110 ft. (27.4–33.5 m) toward the rear of the hall without negative effects. It is more important to have the audience close to the stage in order to add G, envelopment, and intimacy than it is to maintain width in the back portion of the hall. The seats under the balcony will receive sufficient side wall reflections from the throat area, not the wider rear side walls.

Throat Walls

Let us begin with a discussion of the most important wall of all—the throat walls. It is the first third of the side wall, starting at the proscenium, as shown in Figure 12.1. Generally, the walls are angled at approximately 7° off centerline for proper reflections. This will send early reflections to center orchestra seats that need it most for C80 generation. These reflections should arrive as early as possible within the 80-millisecond window. These highly valuable seats in the orchestra are the farthest from any reflective surface and benefit greatly from throat wall reflections, especially if there is a side box or balcony soffit to direct the reflection down into the orchestra seating (see Figure 12.1).

In order to deliver the loudest possible orchestral sound to the audiences, the throat walls gradually flare toward the rear of the hall in a straight, stepped, or gently curved line. Throat walls serve as an extension of the stage area walls when the orchestra or performer is out on the lifts.

These walls can be made of three-layer gypsum board or heavy wood paneling over drywall, but I prefer to use plaster on block or concrete when possible. Excessive mid- and low-frequency absorption is the bane of drywall halls. Having solid, stiff, and massive walls helps to maintain a robust RT and solid bass ratio (BR). If symphony is not a major component of the program, then drywall walls are acceptable. In fact, the 2600-seat Marion Oliver McCaw Hall in Seattle, Washington, has three layers of drywall on all walls and is very successful as an opera and ballet house.

I recommend modest-, small-, or medium-scale diffusion on walls. The desire is for strong early reflections that are not overly diffused and so aid C80 and add energy to the hall. If the diffusion is too aggressive, the throat walls will be less effective and start to be absorptive.

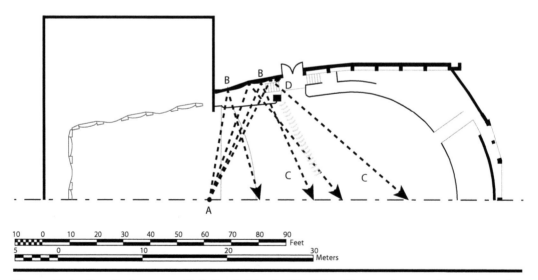

Figure 12.1 **Bass Performance Hall, Forth Worth, TX, 1989. (A) Sound source located at soloist's position with sound rays toward throat walls. (B) Slightly convex sidewall shaping spread. (C) Early reflections to fill in shadow caused by (D) architectural column.**

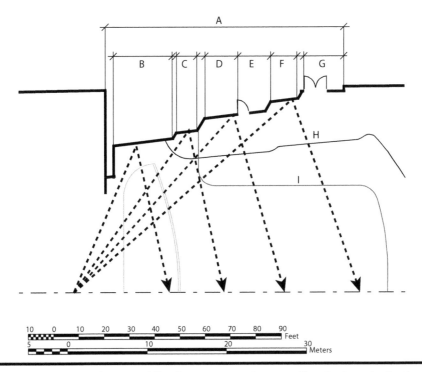

Figure 12.2 **Wagner Noël PAC, Midland, TX, 2009. Throat walls. (A) Zone in the front of the hall for early C80 and D50 reflections. Hold angle of walls (B) through (G) to 7° in steps. (H) Balcony soffits above (I) parterre wall.**

In larger halls, throat walls angled at 7° coming off the proscenium area often render the hall too narrow. In this case, hold the 7° and do not flare the proscenium too widely because that will send all reflections to the rear of the hall. A solution is to simply step the walls, as shown in Figure 12.2.

Stepped Throat Walls

Mid- and low-frequency reflections can be created for the center seating area by stepping the throat walls out in approximately 7-ft. (2.1-m) segments avoiding a wide fan-shaped hall (see Figure 12.2).

These segments step back at about 2–3 ft. (0.6–0.9 m) per segment and are an ideal way to create clarity and excellent C80 levels without having a narrow hall. A very long, narrow hall is pejoratively referred to as a bowling alley hall because it has dimensions so deep that audiences in the rear have no connection to the performance.

The Complexity of Throat Side Walls

The throat side walls, also called forestage, are one of the most complex parts of the hall's design. Many components need integration with this area. Design of this area should commence early in the design process with the architect and the theater consultant.

Theatrical lighting designers jockey for space for lighting positions. Box booms are often recessed into throat walls, requiring that walls be carved open and their reflective quality be compromised. This opening should be kept as small and high as possible.

Loudspeakers (line arrays are now in fashion) can conflict with throat wall shaping. Keep them flown on winches and exposed to view in order to greatly reduce the complexity of throat wall construction. It is a real challenge to effectively enclose speakers in a sound-transparent wrapper. The wrapper often becomes larger than the line array itself and draws attention to itself, defeating its purpose. Upper throat walls can serve as a location for adjustable acoustic drapes that cover throat walls and absorb off-axis sound from side speaker arrays improving speaker performance.

There is often a need for performer entrances on the stage apron, as well as audience entrances to the front of the seating area. These doors need to be thoughtfully incorporated into throat wall angles. Since a proscenium provides a visual focus for the theater, architects often focus on creating an engaging frame for the opening, which further complicates the design of the throat.

Throat walls often incorporate side galleries, boxes, or balconies. This can help acoustically because the soffits below these architectural elements provide additional reflective surfaces (see Figure 12.3).

The double reflection of a soffit coupled with the throat wall is a powerful combination that can drive early wall reflections to very high C80 levels. The soffits need to extend horizontally at least 6–7 ft. (1.8–2.1 m) from the side wall to be effective reflectors. Narrow shelves that extend only 3–4 ft. (0.91–1.21 m) are only partially effective and limit the frequency response drastically. Use three layers of drywall or thick plaster for soffits in order to create full-frequency reflectors. The soffit should be flat, horizontal, and at a 90° angle to the throat wall. There is no value in curving or sloping the soffit. The first balcony soffit provides the most effective reflection in the throat area. Second- or third-tier soffits are less useful but still add value.

Figure 12.3   Schuster PAC, Dayton, OH, 2003. Soffits under the side galleries in combination with the sidewall shaping provides lateral reflections for enhanced C80 to the orchestra seating area.

Side Walls

The area of the wall between the throat wall and the rear wall is also an important surface. This area provides lateral reflections to the seats in the rear of the hall and develops reverberation in the side-to-side energy mode. For this reason, the walls from the back third to the rear of the hall should be mostly parallel to each other with moderate diffusion.

Walls that are parallel develop the strongest reverberant field. Remember, acousticians want a strong and rich reverberant field that decays evenly. The upper side walls must contribute by significantly supporting side-to-side energy. Side walls sustain reverberation and can have gentle convex curves, and modest diffusion elements on side walls foster even blending. The diffusion should have small-, medium-, and large-scale elements to blend at all frequencies.

Rear Wall

Often, architects and theater consultants draw the rear wall as a concave curve that follows the line of the seating. A concave curve has a negative effect on the hall's acoustics and affects artists on stage. The curved surface can send a nasty reflection back to the stage at 150–200 milliseconds, and it is perceived as an echo. This causes a problem with artists hearing themselves twice by blurring overlapped rhythmic lines. If the focal point of the curve is directly in line with the stage, echoes off the back wall can actually exceed the loudness of the sound leaving the stage. Musicians call this disturbing effect a slapback echo. If the wall is concave, speakers aimed at the back wall also create this negative effect.

To control this effect, the rear wall should be treated with diffusion shapes rather than covered with absorption. Better yet, do not shape the rear wall in a concave manner at all. I advocate for the rear wall to be straight across with modest diffusion.

Here is why:

- Entry doors and control room windows cover much of the lower area, and realistically, these surfaces cannot be acoustically treated with diffusion.
- If the rear wall is straight, the sound energy is reflected back toward the stage as an expanding wave front (not a focusing one) and gives the audience a sense of envelopment.
- The rear wall is often under the balcony, and seats in this zone need many sound-reflective surfaces around them in order to maintain loudness and brightness.
- Splaying the rear wall with a series of accordion-shaped pleats or gentle convex curves will spread sound energy from amplified sources. This is an effective way to break up an existing concave rear wall without adding absorption or expensive diffusers. By stepping the wall to conform to the curve, focusing is avoided.

See the shaping detail in Figure 12.1 and read more in the case studies.

Wood Walls

The presence of wood in performance halls is a topic of much discussion and misinformation. I have heard dozens of questionable comments by music professionals and designers in relation to the insistence of wood for walls, ceilings, and floors. Wood is used in flexible performance halls for stage floors and stage extensions, but demanding wood walls and ceilings as a requirement for excellent acoustics is yet another acoustic myth.

Recall that many of the great concert halls of the world have limited wood wall surfaces. Walls in Carnegie Hall, for example, are made of heavy plaster on masonry. Boston Symphony Hall is made of plaster, brick, and steel. Why do some insist on an acoustic need for wood surfaces? Perhaps it is the romantic idea that a hall is like a fine wood instrument, such as a violin or a piano, or that wood is natural and organic when plaster and concrete are not? Some may say that as wood ages and mellows, it adds a rich warm sound.

Those statements are myths. Wood is not a prerequisite for excellent acoustics. Science, experience, and data support that statement. In fact, wood walls can be acoustically problematic if they are shaped poorly and inadequately supported by substructure. The shaping of walls must follow guidelines to obtain the appropriate reflections, reverberation, and diffusion regardless of material. When wood walls are thin, they reduce RT and BR due to low-frequency panel absorption. Wood is also a fire hazard, and its use can be limited by fire codes and regulations.

Psychoacoustics

The use of wood on stage walls surrounding the orchestra is a relatively new phenomenon, more aesthetic than acoustic. It caters to the perception that wood is an excellent reflector for sound. Under the right conditions, this is true. This phenomenon is sometimes referred to as psycho-acoustics or non-acoustic factors that affect artists' and audience's perception of sound. The theory states that a musician may see a wood stage and wood walls and perceive that sound will be better. In fact, the sound measured with electronic instrumentation would be identical if the walls were made of plaster or multilayer drywall.

This may contribute to why wood is a well-liked material for orchestra shells and walls. As discussed in the Alice Tully Hall case study (Appendix I), the wood used is paper thin and mounted on thick resin or adhered to medium-density fiberboard (MDF). This gives it an ability to be dimensionally stable and fire resistive.

I advocate wood for use on surfaces such as balcony fronts, parterres, box railings, and seating where the patron can touch it and feel the wood grain. The use of wood bottoms and wood backs of seats is strongly encouraged and is discussed in more detail in Chapter 14.

Wood Wall Construction

When wood walls are used in large sections, they should be at least 1.5 in. (38 mm) thick and be adequately braced for minimal low-frequency absorption. Consider a large surface to be more than 3 ft. (0.9 m) in any dimension. The theory is that smaller surfaces are not effective as low-frequency absorbers. Therefore, relaxing the thickness criteria for cost savings becomes possible (see Figure 12.4).

Wood veneers on a solid substrate such as MDF are common. Yet, attaching the MDF to solid masonry or concrete in a manner that does create panel absorption is complex. Never use Z-clips or hat channels as this spring loads the panels and accentuates low-frequency absorption especially below 200 Hz. I advocate for these wall clips when the acoustic design calls for extensive use of panel absorbers to reduce low-frequency reverberation. This may be the case for a hall that is used for all-amplified events. Multi-use halls used for classical music and opera should avoid using wood panels on clips. MDF panels can be adhered with construction adhesive and fasteners directly to a massive substrate such as masonry or multilayer drywall that lacks air space.

Figure 12.4   Baldwin Auditorium, Duke University, Durham, NC, renovated 2013. Convex-shaped and veneered wood walls with added articulation from applied wood elements form an ideal stage enclosure. Baldwin Auditorium is featured in a case study.

Drywall Walls

Solid masonry, thick plaster, and solid wood walls are always an acoustician's first choice; however, budget or schedule often mandates the use of the drywall on metal framing solution. There are significant cost advantages to this approach, and excellent results can be obtained for halls with fewer than 1500 seats.

As with wood, the danger of excessive panel absorption is always present. Drywall areas should be carefully balanced with solid masonry such as the ceiling, roof, or floor system. Never use single-layer drywall construction, and limit two layers to small areas of 10% or less of wall surfaces. Three layers or more are required for an unyielding surface that successfully reflects low frequencies (see Figure 12.5).

To render walls more inflexible, tighter stud spaces can be used. For example, use 12 in. (30 cm) rather than the typical 16 in. (41 cm), but be sure to vary stud spacing or use structural gauge studs.

I do not advocate gluing the layers together. Typical drywall multilayer construction as detailed in the manufacturer's handbooks is adequate. Drywall taping and finishing should be used on the outer layer only.

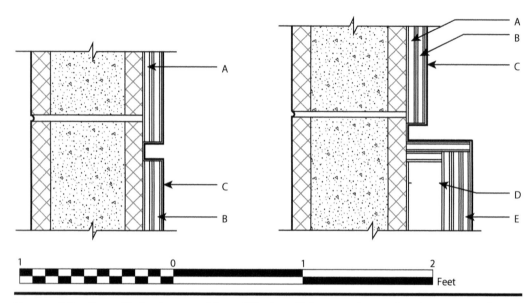

Figure 12.5 Drywall shaping to achieve diffusion. This example, used at Wagner Noël, uses built-up layers of drywall to reduce LF absorption. (A) First layer laminated to drywall with nails and glued. (B) Additional drywall layer mechanically attached to form rigid surface. (C) Skim coat plaster finish. (D) Metal studs attached to block walls. (E) Three layers of drywall attached to metal studs with skim plaster finish.

Masonry Walls

Rigid, stiff walls are vital to obtaining the proper acoustic response in a hall. Walls above the visual ceiling can be masonry or block but must be thoroughly painted and sealed with block sealer. The Wagner Noël Performing Arts Center case study (Appendix X) has more information on this topic.

Block walls come in a plethora of thicknesses, weights, and densities. I had to learn the hard way that all concrete blocks are not alike. Very coarse and lightweight block is often less expensive than heavyweight block. This is sometimes referred to as normal-weight block. In the US, there are different standards for weight and density, and the acoustician must be adamant about using heavyweight block with the highest possible density.

Plaster can be applied on block with or without a metal mesh or metal lath. I have applied stucco plaster directly to block for a very firm and inflexible surface. Wetting the block with water before the stucco is applied allows for a monolithic, solid surface. If lath is used, the plaster must adhere to the block so that there are no air gaps to foster low-frequency absorption.

Painting block is not a simple process for a sealed and closed surface with minimal absorption. First, the block sealer must be rolled on and then covered with many more layers of thick, rolled-on (not sprayed) paint that seals the pores of the block completely. A closeup inspection with a powerful flashlight will reveal if the block is acoustically rock hard. Unpainted pores look like white dots under a strong light, and additional layers will be needed to cover them.

Concrete Walls

While not considered an acoustic material per se, poured-in-place (PIP) concrete and precast concrete should be considered an acoustic tool. Rock hard and unyielding, PIP is an excellent surface for specular reflections and for maintaining low-frequency sound energy. With its obvious transmission loss attributes, PIP's only downside is its cost and the lack of ease of casting sound diffusion into the forms. At the Dallas City Performance Hall, walls are PIP and are parallel to each other. Apprehension with flutter echoes and acoustic glare led to the use of board-formed concrete finishes. These finishes articulate the surface with horizontal bands of textured concrete that produce mid- and high-frequency diffusion. The board offsets reassured me that adequate acoustic diffusion was achieved. Additional diffusion was contributed by the overhead reflector, catwalks, and balcony structures. Please see the case study on the Dallas City Performance Hall (Appendix IV) for more information (see Figure 12.6).

Figure 12.6   Dallas City Performance Hall, Dallas, TX, 2012. Concrete wall. While not typically considered an acoustic material in hall design, it can be effective when used correctly.

Precast surfaces of a similar design were used on the side walls at the Joan W. and Irving B. Harris Theater for Music and Dance in Chicago, Illinois. The hall was built as part of a huge multilevel underground precast concrete parking garage under Millennium Park on Michigan Avenue. Precast, 12-in. thick walls (30.4 cm) were complete with diffusive surfaces as aggressive as the manufacturer would permit. While modest in depth, these walls met the diffusion, isolation, and reverberation requirements.

Be wary of having all concrete surfaces in a multi-use hall. Massive surfaces might cause excessive buildup of low-frequency reverberation and excessive BR. Balance the reverberation with two-layer drywall or plaster surfaces on ceilings, soffits, or acoustic reflectors.

Myths and Misconceptions

Myth #1: Halls with Wood Walls Have the Best Acoustics

Wood walls can be problematic for acoustics if shaped poorly and inadequately supported by substructure. Carnegie Hall, Boston Symphony Hall, Wiener Musikverein, and the Concertgebouw all have thick plaster walls, not wood.

Myth #2: Concrete and Block Walls Make for a Harsh Acoustic

Not true. I began this chapter with a warning that the devil is in the details. These materials can achieve excellent acoustics when carefully shaped, diffused, and constructed.

Myth #3: Drywall and Plasterboard Are the Best Acoustic Materials for Walls

Not always. I have used three layers to comply with budget restraints in a large hall but would much prefer plaster, sealed block, or concrete in order to achieve a strong BR.

Myth #4: Concrete Block Is a Sound-Reflective Material Useful for Hall Walls

Not true. Concrete block can be used only if fully sealed with block sealer and many coats of bridging paint. Raw block has an absorption coefficient of 0.2–0.3. Hint: paint a sample and see if water will pool for a sustained period of time (see Figure 12.7).

Figure 12.7 **Concrete masonry unit (CMU) painted. When properly sealed (waterproofed), CMU can be an effective material for multi-use halls. Note how the water pools on the painted CMU.**

Chapter 13

Ceiling Designs

Introduction

There are many acoustically successful ways to design the ceiling of a multi-use hall. The ceiling and roof proposal will need to address many competing goals. For example, there must be adequate space for adjustable acoustic banners, drapes, or panels to be stored and deployed to control RT and overall acoustic energy. Diffusion must be incorporated into the design as well as surfaces for reflections to enhance clarity and control echoes and focusing.

The ceiling provides space and support for AV systems including main reinforcement speakers, normally located near the stage and over balcony units flown further back. Theatrical lighting positions must also be incorporated, as well as house lights, work lights, fire sprinkler systems, and smoke detection devices. There must be room for adequate acoustic isolation to protect from environmental noise, including rain impact.

Given that the floor plans define the seating, sightlines, and circulation, ceiling height must be necessarily defined by the acoustic volume and reflection criteria.

Where to start with the design of the ceiling in a multi-use hall? What should the acoustician be thinking about as he/she guides the design team toward a solution?

Acoustic Strategy

The ceiling offers many acoustic opportunities and challenges and may be the most complex element of the multi-use hall design. The forestage reflector zone is particularly problematic because there is a multitude of divergent and opposing requirements from the acoustician, the theater consultant, the architect, and engineers. When there is frustration with this conflicting criteria, I often remind the architects that this is always the most complex area of the hall. We always find a solution but rarely on the first pass.

Ceiling Approaches

There are three types of ceiling approaches. Each is a viable and successful approach that presents options and challenges to the practitioner. When construction cost is a key driver, the open and sound-transparent plans have an advantage because these options supply the needed acoustic volume with a more compact building envelope and lower exterior walls. The design architect may not support this aesthetic.

With this option, the roof alone must furnish the sound-reflective massive reflector for RT support at all frequencies, as well as the sound isolation and rain noise reduction. Any exposed duct work or other exposed systems in the upper volume must not be neglected because it may affect the mean free path. Spray-on fireproofing on structural steel, must be avoided as it can be an unwanted sound absorber.

Finished Ceilings

A finished ceiling is usually made of plaster, wood, or multilayer drywall and encloses the upper volume of the hall. To achieve the volume criteria, the ceiling ends up being quite high. This forces the catwalks to be distant and the adjustable acoustic systems to be exposed on walls. Acoustic banners become more complex also, and storing them becomes more detailed because they are in plain view (see Figure 13.1).

Sound isolation through the roof is improved by having the large air cavity. Ducts, pipes, and conduit are hidden and help to achieve a finished look to the hall. The ceiling is costly and difficult to construct but allows for acoustic reflections and diffusion materials to be holistically integrated.

Alice Tully Hall in New York, Daegu Opera House in South Korea, and Bass Hall in Fort Worth, Texas are excellent examples of halls with finished ceilings and are featured in case studies in Appendices I, V, and VIII, respectively (see Figure 13.2).

Open Ceilings

An open or exposed structure ceiling is often driven by a tight budget. All the ducts, pipes, and structural systems are painted a dark color so that they disappear from view. Acoustically, this type of ceiling allows the roof to be lower to achieve the volume criteria since the concrete underside of the roof now defines the volume. However, the roof and the upper side walls must provide all sound isolation from the exterior.

Exposed elements must be factored into acoustic calculations because they can reduce the mean free path and add absorption and diffusion. The advantages of using an open ceiling are lower initial cost, ease of construction, ease in operation of adjustable acoustic systems, lower roof height, and more flexible catwalks and forestage rigging. Acoustic reflectors are necessary in order to provide appropriate overhead reflections and diffusion elements. Important design factors include open-to-closed reflector areas, size of the elements, and construction materials (see Figure 13.3).

More examples can be seen in the case studies for the Michael and Susan Dell Hall at the Long Center for the Performing Arts, Texas (Appendix VII) and The Wagner Noël Performing Arts Center, Texas (Appendix X).

Sound-Transparent Ceilings

A hybrid of the two ceiling options creates a third option, a sound-transparent ceiling that is made of a visually opaque yet sound-transparent material. The ceiling looks finished and solid but allows sound to freely move through it. This treatment includes the cost benefits of having a lower roof height and having less complex locations to place adjustable acoustic banners and reflectors. Rigging and catwalks are more easily accommodated than with the fixed ceiling design. The upper walls and

Figure 13.1 Bass Performance Hall, Forth Worth, TX, 1998. An example of a finished ceiling design for a multi-use hall. There are minimal openings for lighting and air distribution. The large openings have doors that close to cover the openings in performance mode.

the roof must provide the required exterior sound isolation without the assistance of an inner layer of finished ceiling. The acoustic transparency of a mesh, wood slat, or perforated metal ceiling is not a trivial issue. It must be sound transparent from both normal and oblique angles and exhibit minimal losses, discoloration, or distortion. This type of ceiling is utilized in the Benjamin and Marian Schuster Performing Arts Center, Ohio, and the Wallis Annenberg Center, California, and are featured as case studies in Appendices III and XI, respectively (see Figures 13.4 and 13.5).

Fixed Forestage Ceiling

For the design directions of a movable forestage that is part of the orchestra shell, refer to Chapter 10. The zone with the most complexity in a multi-use hall is the first 10–20 ft. (3–6 m)

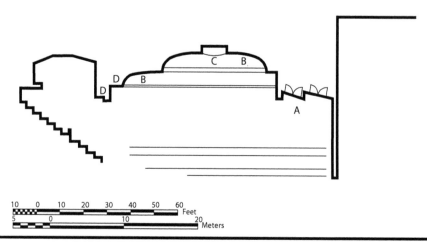

Figure 13.2 Bass Performance Hall, Forth Worth, TX, 1998. Finished ceiling design. (A) Forestage ceiling angled for proper pit reflections and operable to allow forestage reflector and rigging to drop through (B) double dome with constant varying radii. (C) Glass light fixture. (D) Theatrical lighting positions.

in front of the proscenium. The side walls and the ceiling area perform myriad complex acoustic, theatrical, and architectural functions that change depending on the performance type and genre. I advise allowing flexibility in this zone so that future technologies for video, audio, and lighting can be accommodated. The acoustician should advocate that forestage walls and ceilings must support unamplified musicians, singers, and actors with early reflections that increase G and C80.

Height

The height of the fixed forestage reflector should be no more than 40 ft. (9.1 m) above the stage to deliver strong and appropriately timed reflections. Movable forestage reflections are discussed in detail in Chapter 10.

Halls with very strong lateral reflections from narrow walls of perhaps 60–70 ft. (18–21 m) and soffit reflections do not necessarily need a reflector. I do endorse a reflector for academic spaces because students need every possible advantage to bolster their confidence and improve hearing.

I also advocate for use of a fixed reflector to provide additional clarity, loudness, and articulation for classical music performance. This is my preference because I feel that a hall should provide a truthful, unforgiving sound rather than a thick, nebulous one. The forestage is a vital component that reinforces that desire.

The forestage ceiling also allows sound energy to pass freely into the reverberant zone above to provide strong RT development. A forestage piece that is overly large will block too much energy from reaching this upper volume. I suggest that the reflector extend no further out in the hall than the lift below and have strategically placed openings.

Create a gap at the proscenium of 2–3 ft. (0.6–0.9 m) to accommodate forestage lighting and to allow sound to pass. Also, leave gaps up and down stage to allow for rigging lines from the forestage grid and to further vent the forestage reflector. The ideal reflector will consist of three

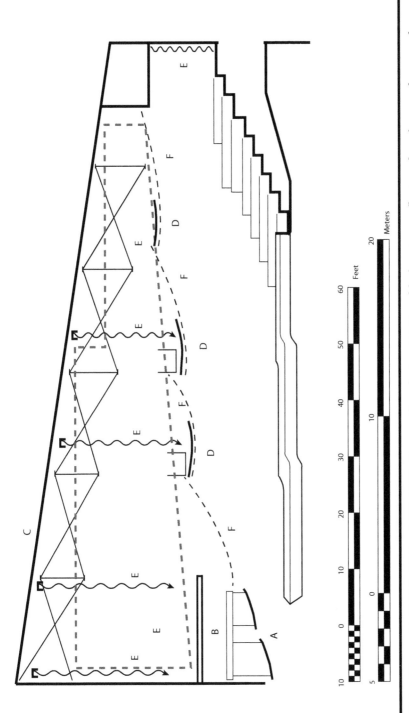

Figure 13.3 Wagner Noël PAC, Midland, TX, 2009. Open ceiling design. (A) Movable forestage reflectors drop for symphony performances. (B) Forestage grid for rigging. (C) Concrete roof on metal decking on exposed metal trusses. (D) Wood bent acoustic reflectors. (E) Adjustable acoustic banners and drapes. (F) Suspended LED light curtain, open acoustically.

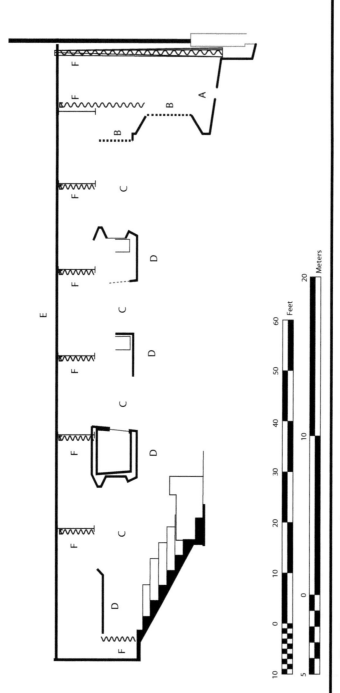

Figure 13.4 Dell Hall at Long CPA, Austin, TX, 2008. Sound transparent ceiling. Partial ceiling shown at (A) forestage reflector with rigging and lighting openings and (D) three-layer drywall ceiling elements under catwalks and spot booth, all more than 50% open. (C) Open areas to upper volume. (B) Vertical sound transparent grilles. (E) Concrete roof on metal deck on exposed steel trusses. (F) Adjustable acoustic drapes.

Figure 13.5 Schuster PAC, Dayton, OH, 2003. Sound transparent ceiling. Metal mesh ceiling appears solid when viewed from below. Clever lighting conceals the very open mesh.

layers of drywall or 1.5-in. (3.8-cm) wood or plywood. However, if budget does not allow for this, lighter-weight reflectors can be used with minimal loss of sound energy. I have used a limited amount of 0.75-in. (1.9 cm) material with no ill effect. The Michael and Susan Dell Hall at the Long Center for the Performing Arts, Texas, case study (Appendix VII) shows a fixed forestage reflector that follows the design criteria.

Shaping

Shaping the reflector can make a huge difference. I advise shaping a gentle curve to the face of the forestage so that ceiling reflections are evenly distributed, and onstage hearing is improved. In this way, the reflector serves as an extension of the orchestra shell ceiling forward of the proscenium.

I advise the ceiling piece be shaped in a complex double curve if the side wall reflections in the forestage area are bereft. My firm utilized this technique with the Baldwin Auditorium's forestage at Duke University for precisely that reason and is visible in the case study images in Appendix II and in Figures 13.6 and 13.7.

A smooth, nondiffusive surface is usually acceptable except when some left and right reflections are needed. The side walls in the forestage are far more effective for reflections across the stage to promote onstage hearing.

The acoustician should collaborate with the AV designer on speaker cluster access around or through the reflector. I have found that it is not necessary to create a complex plug for the small opening for the center array or side arrays to pass through.

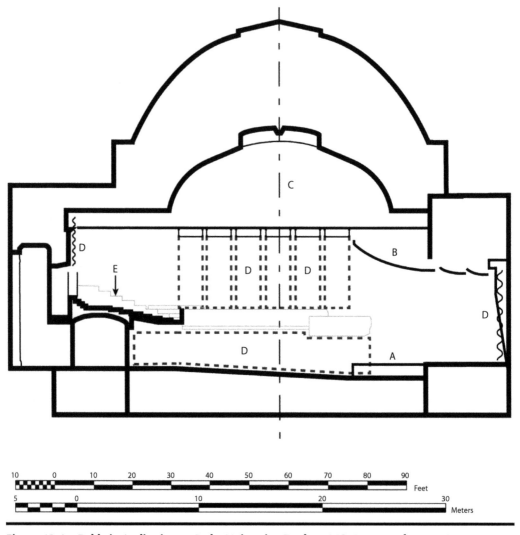

Figure 13.6 Baldwin Auditorium at Duke University, Durham NC, Renovated 2013. (A) Permanent stage extension. (B) Wood acoustic canopy reflector. (C) Existing dome ceiling of thin plaster. (D) Acoustic drapes and banners. (E) Balcony overbuild.

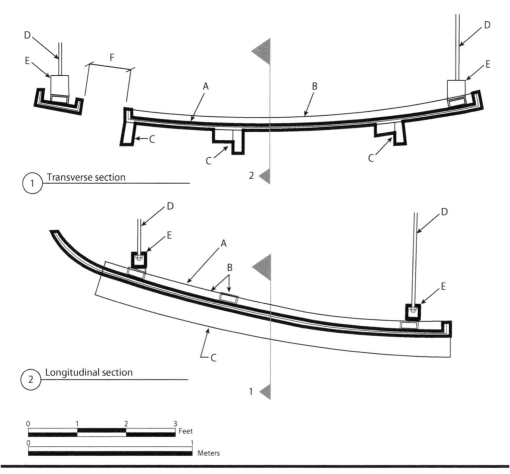

Figure 13.7 Baldwin Auditorium at Duke University, Durham, NC, renovated 2013. Double curved ceiling reflectors. (A) 1 inch wood substructure. (B) Wood support ribs. (C) Applied sound diffusion elements. (D) Threaded rood support system. (E) Metal clip to attach rod to wood ribs.

Myths and Misconceptions

Myth #1: A Closed Ceiling Is Superior in Terms of Isolating Exterior Noise

While this is usually the case, I have achieved excellent isolation of an open ceiling design by using a floating 4-in. (100-mm) concrete slab on neoprene isolators placed on top of the concrete roof slab. This provides the benefits of an open ceiling and superior sound isolation.

Myth #2: Avoid Dome-Shaped Ceilings

Concave ceilings can be problematic but can also be an asset if the appropriate diffusive elements are applied and if the dome has a variable radius rather than a constant radius. I have found that it is possible to send positive reflections to performers from the dome. This is called return and is especially helpful to opera singers. Nancy Lee and Perry R. Bass Performance Hall in Fort Worth, Texas, has such treatment.

Myth #3: The Ceiling Reflector Should Send Reflections to the Back of the Hall to Increase G Levels

This is not true. If correctly designed, the rear of the hall should receive lateral reflections from the side walls and the ceiling area. Forestage ceiling reflectors should direct reflections to the center orchestra level to increase C80 and G levels.

Myth #4: Ceiling Openings for Theatrical Lighting Should Be Plugged with Solid Doors or Glass Panels When in Symphony Mode to Keep Sound from Leaking

False. The openings are so small in relation to the overall ceiling area that the loss in sound is negligible. My experience is that technical crews rarely close the doors after they are opened.

Chapter 14

Seats and Finishes

Introduction

An architect designing his first multi-use hall once remarked to me in exasperation, "Acousticians have an opinion about every material, surface, and construction in the hall. Is there *anything* that you don't weigh-in on?" I explained that all materials in a multi-use hall are acoustical in nature, from the fabric on the seats to the pile height of the carpet. In fact, acousticians have a recommendation on everything but the color. I quipped, "You can choose that; just don't choose blue." It is attention to detail that takes a good hall and makes it excellent.

Seating

Seating elements and seat configuration are a significant acoustic component of a hall and have a major influence on its acoustics. In a multi-use-seat hall, chairs can represent 80%–90% of all nonadjustable absorption. Differences in seat absorption will have a profound impact on reverberation time (RT), G levels, and bass response.

Accommodating Patrons

Seating plans are driven by the theater consultant and architects' concern for good sightlines, appropriate comfort, and adherence to building code criteria for emergency egress. Added complexity comes from the Americans with Disabilities Act, a law that defines critical criteria for access to seating areas for patrons with limited mobility. The United States has some of the strictest criteria; however, other countries are instituting similar criteria to provide audience members with access to most seating areas throughout the hall including the front rows, side boxes, and rear balcony. It is important to note that a wheelchair has a larger footprint than a typical chair. Thus, the seating density is decreased, and RT is reduced.

There has been a trend over the last decade to increase the width of seats. In the 1990s throughout the 2000s, 18–20 in. (460–510 mm) wide seats were typical. However, now the preferred seat width is 20–22 in. (510–560 mm) in the United States.

The distance between rows, or row-to-row depth, has also grown since the 1990s to provide extra leg room and meet new building code criteria for egress. Increased row-to-row depth means that the gross floor area required for a given audience capacity has grown considerably, and there is now additional sound absorption for the fixed seat count.

Modeling Variances

The acoustician models the acoustics of a multi-use hall with seats occupied and unoccupied in order to determine the acoustic response for rehearsals and lightly attended events. The hall will always be more reverberant when it is empty. Occupied seats are obviously more absorptive than unoccupied seats; however, the geometry of the hall, the seating slope, and seat construction will influence variances in RT, G, and C80, respectively. Steeply raked seating areas and halls where the audience surrounds the stage, also known as vineyard halls, exhibit increased empty to occupied variances.

Seat Dip Effect

Another factor to consider is the seat dip effect. This effect occurs in most halls and is explained as the reduction of sound at grazing incidence as it passes over the main floor seating section.

The main floor of any hall can be characterized by a weakness in certain sound frequencies. This frequency range spans from G (100 Hz) at the bottom of the bass clef to D (500 Hz) at the top of the treble clef. It covers the primary range of violoncellos, violas, trombones, bassoons, and the bass and baritone voices.

For low-frequency sound, the attenuation is in excess of the inverse-square loss due to distance, typically 15–20 dB. This defect is most apparent at approximately 150 Hz. The defect gradually worsens from the first row of seating through about the twelfth row, at which point degradation levels off.

The seat dip effect is an interference effect and does not depend on the absorptive properties of the seats. The frequency of maximum attenuation is related primarily to the height of the seats and only secondarily to the row-to-row spacing. The dip in transmission for the seats occurs at about the frequency where the seat height is a quarter of the wavelength. Therefore, an average concert hall seat height of 2.5 ft. (0.75 m) would yield a seat dip effect centered at approximately 113 Hz. Extending the seat back down until it intersects the floor has been suggested as a way to reduce seat dip, but it has yet to be tested in the real world. The acoustician must be aware of the seat dip effect and its effect on the hall.

Guidelines for Seats

Seats should have the least possible amount of sound absorption no matter if they are occupied or unoccupied. Seats must also meet basic comfort criteria. I advocate the use of adjustable acoustic systems such as banners and drapes to make an empty hall's RT more similar to the occupied mode. I have found this to be more successful than making the empty seat more absorptive through the use of seat materials such as perforations.

I have found that it is not worthwhile to perforate the bottom of seats. The absorption introduced by perforations exists in the hall whether or not the chair is occupied. Historically, perforations in a metal seating pan were used to add absorption by the function of the contained air volume. Again, this absorption existed whether or not the chair was occupied. A wood bottom lacks the air volume, so perforations are of no value and only add cost.

AUDIENCE/SEAT ABSORPTION	63 Hz	125 Hz	250 Hz	500 Hz	1000 Hz	2000 Hz	4000 Hz
Occupied, heavily upholstered	0.50	0.72	0.80	0.86	0.89	0.90	0.90
Occupied, medium upholstered	0.43	0.62	0.72	0.80	0.83	0.84	0.85
Occupied, lightly upholstered	0.36	0.51	0.64	0.75	0.80	0.82	0.83
Unoccupied, heavily upholstered	0.49	0.70	0.76	0.81	0.84	0.84	0.81
Unoccupied, medium upholstered	0.38	0.54	0.62	0.68	0.70	0.68	0.66
Unoccupied, lightly upholstered	0.25	0.36	0.47	0.57	0.62	0.62	0.60

Figure 14.1 Seating absorption coefficients.

Seat Absorption

Research from Leo Beranek and others details three categories of seat absorption: light upholstery, medium upholstery, and heavy upholstery. Multi-use halls primarily utilize chairs with medium-upholstery acoustic properties in order to minimize seat absorption and provide comfort for audiences sitting for a three-hour performance. My experience is that heavily upholstered seats are actually no more comfortable over the duration of a performance than a firm, well-designed foam seat of modest thickness. The data are accurate in modeling RT within 5%–10% accuracy (see Figure 14.1).

Beranek's Seat Absorption Method

To calculate the absorption of the audience area, count the seats and a half-meter zone (1 ft. 7 in.) around the bank of chairs. Do not add a zone for chairs that share a boundary with balcony rails or walls.

Recommendations on Seating Materials

These data are based upon acoustically successful seats in real halls, not just from a small number of seats tested in a lab. I put a high value on experience-based design and use laboratory tests to augment and expand my knowledge. I recommend using wood materials for seat backs, bottoms, armrests, and end standards. The wood thickness should be ¾-in. (19 mm) or greater with a high-laminate plywood with hardwood veneer. We recommend a wood surround of 1–2 in. (25–50 mm) on the seat back to reduce the area of exposed upholstery to the bare minimum. Fabrics add absorption to the hall.

Plastic and metal bottom chairs are discouraged, except when the budget is very limited, because the seat cavity resonance adds undesirable bass absorption. It is vital that the seat back foam be kept to the minimum thickness acceptable for comfort, usually 1.5 in. (38 mm) expanding to 2 in. (50 mm) or slightly more in the lumbar area.

Seat bottom foam, likewise, should be kept to a thickness of 2 in. (50 mm) and be directly adhered to the wood bottom. Foam density has a modest acoustic impact. The higher-density foams are generally less absorptive, but the results are inconsistent chair to chair. The thickness is more significant. Woven fabrics, velvets, and mohair are typically more absorptive than leather, perforated, or automotive seating materials. The seats in Alice Tully Hall at the Lincoln Center for Performing Arts in New York City are made from materials found in high-end German luxury automobiles (see Figures 14.2 through 14.4).

Figure 14.2 Theater seat with pedestal air diffuser. Note the thin foam back with exposed wood reveal and thin foam bottom on wood.

Figure 14.3 Custom theater seat. Thin foam back and bottom are acoustically effective.

Avoid internal circulation stairs, overly wide aisles, and internal corridors within the acoustic envelope as they tend to spread the audience out and make the hall larger than necessary. Seats should be packed closely together as much as comfort and egress will allow. There is no acoustic value in knee walls, or pony walls, being placed parallel to the seating rows.

Loose chairs in side seating and in boxes are a small yet significant element that the acoustician should not forget about. Follow the guidelines of using thin foams and plenty of wood materials. Lightweight chairs should be considered because they are compact, easily moved within the box, and modest in footprint. Loose chairs can scrape on a noncarpeted floor in box tiers and produce noise disturbances during performances. To prevent such intrusion, treat the chair bottoms with guides that smooth the vibration of the chair leg. Acousticians prefer not to carpet areas under loose seats. Teflon-based glides such as Magic Sliders fasten to the bottom of the seat legs and are successful at mitigating noise.

Figure 14.4   Schuster PAC, OH, 2003. Loose seats in boxes must also meet all acoustic criteria. Use guides on seat legs to reduce noise from scraping on hard floor.

Finally, a certain amount of noise is inevitable when the patron stands up and the seat rises to a vertical position. This noise can be obtrusive unless the mechanism is designed to be quiet. Gravity-operated seat operators are quieter than spring mechanisms. Bumpers and dampers help to quiet the seat when it returns to an upright position.

Town Hall's Landmarked Seats

In the 1980s, Jaffe Holden was part of a team commissioned to undertake the renovation of Town Hall in New York City, originally designed in 1919 by the architectural firm of McKim, Mead & White. Built as a hall for the lively discussion of important issues of the day, the acoustics were very well regarded and favorably compared to Carnegie Hall.

In 1921, German composer Richard Strauss gave a series of concerts that cemented the hall's reputation as an ideal space for musical performances. Famed violinist Isaac Stern made his New York debut in Town Hall in 1937.

During this modest renovation, it was critical to preserve the acoustics of the hall. Stage modifications to improve egress and lighting were deemed to have limited impact on room acoustics; however, all 1500 seats were in desperate need of replacement. Jaffe Holden's plan was for the new seats to match the acoustic absorption coefficients of the old seats. The original seats from 1919 were sent to a lab for testing and forensic analysis. To my surprise, the seat absorption coefficients came back with remarkably unexpected results of very minimal absorption in all but two octave bands.

The lab described the seats' unusual composition of mohair fabric over horsehair padding on top of rubberized palm fronds. Open- and closed-cell foams were laboriously tested to match antique materials in an attempt to preserve the landmarked construction of the seats. By evaluating dozens of chairs with various amounts of foam, thickness, and density, we determined the best possible acoustic match for new chairs. However, this chair was horribly uncomfortable. In the end, the client agreed to select a comfortable chair with the lowest possible acoustic absorption, and the hall has continued to receive exceptional acoustic reviews.

Floor Materials

The acoustician designing the acoustics of a multi-use hall advocates for materials and surfaces that sustain low frequencies and BR and maintains this despite budget pressures. The floor construction must avoid excessive low-frequency absorption as well as undesirable high-frequency absorption that would occur from carpeting materials covering areas under seating.

Floor materials that excel include painted or stained concrete, vinyl tiles or linoleum, and wood flooring that is attached to the concrete subfloor leaving minimal air space. Figure 14.5 depicts that the construction of the wood floor in the aisles area was carefully designed to minimize low-frequency absorption.

A special floor construction is possible in renovations where the weight of a concrete floor in a balcony would be structurally excessive and therefore unfeasible. This floor construction is made of stiff but lightweight materials to minimize low-frequency absorption. This type of floor construction has been successfully instituted in balconies of historic hall renovations through the use of an overbuild technique. This is when the new floor is built over the existing historic concrete floor. The new floor solves poor site line issues, increases row-to-row depth, and solves egress issues without having to demolish the existing floor and add the weight, cost, and time to lay a new concrete floor (see Figure 14.6). The Richmond CenterStage case study (Appendix IX) discusses this floor as well.

Carpeting

Carpeting in a multi-use hall is a subject of much discussion among donors, operators, and the design team. Donors desire the plush, rich look that expensive carpet can portray. Operators find carpet to be a maintenance headache because it can be difficult to clean and requires regular replacement. The architect would rather not use carpet in a contemporary building. The acoustician utilizes carpet to reduce footfall and noise from hard and high-heeled shoes when latecomers enter the hall. These competing desires are brought to the attention of the acoustician, and at times, we must come up with a solution that meets all the needs of stakeholders.

Every multi-use hall design is unique, but a standard solution is to have carpet in the main aisles, cross aisles, and minimally in circulation areas. Never carpet areas under the fixed seats. Only carpet the side boxes if absolutely necessary. Many donors insist on carpet, so try to use the thinnest carpet you can find, and do not use a backing pad. Both Dallas City Performance Hall and Alice Tully Hall have wood aisles, and latecomer footfall noise is at times audible.

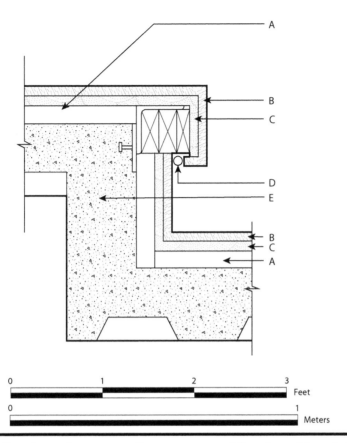

Figure 14.5 Dallas City Performance Hall, TX, 2012. Wood floor in multi-use hall seating area. (A) Wood sleeper systems directly attached to concrete floor. (B) Oak finish T&G wood floor system. (C) Plywood subfloor system. (D) LED step light. (E) CIP concrete floor system.

Sound-Transparent Materials

Sound-transparent materials include grilles, meshes, slats, gratings, and perforated metals. These items are used by the acoustician to provide an architectural cover to an acoustic volume like the walls or ceiling. They also provide an alternate shape for acoustic materials for aesthetic reasons and lower the apparent ceiling to make the room feel more intimate. These materials can be used to cover the ceiling of a sound-transparent balcony, to cover the proscenium, to obscure loudspeakers on walls, to cover acoustic drapes and banners on ceilings or walls, and to cover and protect fixed panels of acoustic treatment.

There are hundreds of acceptable surfaces available to the acoustician and the design team. However, the acoustician must be wary of materials that may look open and transparent but in fact produce coloration or sound distortion that would be noticeable in the reverberant field. In general, a sound-transparent material over fixed acoustic absorption panels like glass fiber requires less critical criteria for openness and sound transparency than a material that covers loudspeakers or a material stretched between two volumes that are acoustically connected.

When materials are acoustically critical, and sound must pass through uncolored in frequency, it is critical to evaluate the percentage of openness and sound. In dry air of 68°F

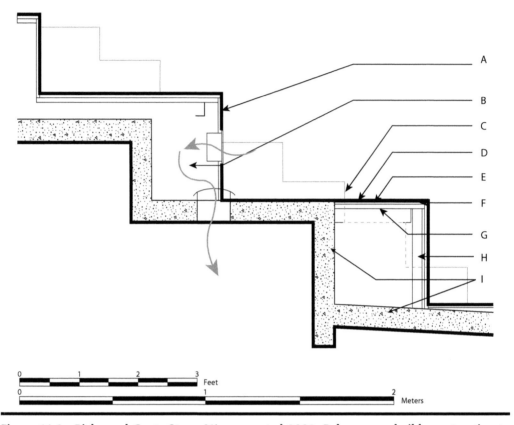

Figure 14.6 Richmond CenterStage, VA, renovated 2009. Balcony overbuild construction to improve sightlines and acoustics. (A) Drywall with plaster finish. (B) Return air plenum with new grille in face. (C) Carpeted steps beyond. (D) Linoleum finish on seating risers. (E) 3/4 in. plywood (2 mm). (F) Damping sheet. (G) 3/4 in. (2 mm) fireproof decking. (H) Metal stud framing 12 in. (0.3 m) O.C. (I) Existing concrete floor and risers.

(20°C), the speed of sound is 1125 ft./second (343 m/second). In a multi-use hall with an RT of 2.0 seconds, sound travels at least 2250 ft. (686 m) before becoming inaudible. A sound-transparent ceiling or wall material might be impacted 20 or more times during the 2-second decay period. Any coloration or distortion is amplified by the number of passes through the material.

I have compared this to the visual effect of multiple layers of what appears to be very clear glass in fact changing the color of the image to a green tint. Imagine a sound-transparent material that has 3-dB attenuation at 5000 Hz. In addition to the natural decay, the sound energy of 20 intersections could be reduced by as much as 60 dB. Materials that are required to be totally transparent must be examined not only at normal incidence perpendicular to the surface but also at oblique angles as well.

This formula was developed by Schultz (1986), of Bolt Beranek and Newman, for sound transparency of perforated materials and is useful for a first pass at material selection. However, the professional acoustician must be absolutely certain that the material will not adversely color or attenuate especially if used in a large area.

Schultz Formula

$$\text{Transmission index} = \frac{nd^2}{ta^2} = \frac{0.04P}{\pi ta^2}$$

where

$n =$ number of perforations per square inch
$d =$ perforation diameter in inches
$t =$ sheet thickness in inches
$a =$ shortest distance between holes in inches; $a = b-d$ where
$b =$ on-center hole spacing in inches
$P =$ percent (not fractional) open area of sheet

An approximation for the value of "a" when you do not know the value of "b" is

$$a = d\left[\left(\frac{\text{constant}}{P^{1/2}}\right) - 1\right]$$

where the value of the constant is 9.5 for staggered and 8.9 for straight perforations.

The attenuation at 10 kHz is thus

$$A = -22.56 \log \log TI + 0.008\sqrt{TI} + 13.79.$$

Rules of Thumb

Always use the material with the least depth, or thickness, to reduce oblique incidence attenuation and coloration. Perforated metals and mesh make the best surfaces.

Wood Slats

Wood slats, also known as wood grilles, are available from a number of manufacturers including Ventwood and Rulon. In order to be truly sound transparent, these materials must be very open and have minimal depth to reduce non-normal incident attenuation. I prefer round sections and doweled supports to create more open area (see Figure 14.7).

Perforated and Microperforated Wood

Perforated and microperforated wood are acceptable for some absorption purposes but are not open enough for sound-transparent surfaces. Very large perforations 3–5 in. (75–125 mm) in diameter in plywood can be used in noncritical applications. Beware of off-axis attenuation and reflections from the solid material between the perforations (see Figure 14.8).

Metals

Metals form the most common and most useful range of sound-transparent materials.

Figure 14.7 **Example of a wood slat system, 70% open, 5/8 in. width × 2 1/4 in. depth slats, 2 in O.C. Use judiciously as off-axis sound transmission is compromised.**

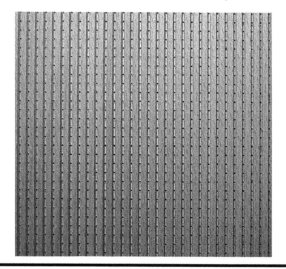

Figure 14.8 **Micro perforated wood, 4% open, 0.95 NRC, A mount 2-in.-thick core. A useful sound absorbing material, but not sound transparent.**

Perforated Metal

The Schultz equation helps to identify a large number of materials. Be wary of the metal framing behind the material and the possibility of buzzes and rattles. Very large openings can be useful for sound transparency (see Figure 14.9).

Expanded Metals

These are excellent sound-transparent materials and tend to look more solid even with very large percentage openings. Expanded metals are more transparent than perforated metal for non-normal incident sound (see Figure 14.10).

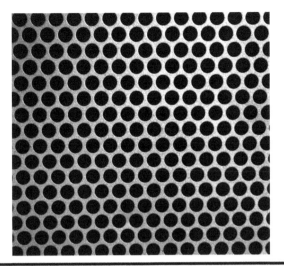

Figure 14.9 Sound transparent perforated metal, 1/4 in. round on 5/16 in. staggered CTRS, 58% open. Considered sound transparent in most cases.

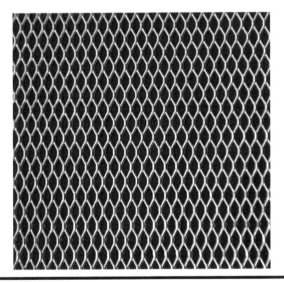

Figure 14.10 Expanded steel mesh, 66% open. Very sound transparent and excellent off-axis transmission.

Grates

Industrial grating materials are expensive but can be used as sound-transparent wall materials. Care needs to be taken for off-axis attenuation by the depth of the supporting system behind the face grating.

Woven Metal Mesh

This is a fine sound-transparent material as it is metal wire that has little depth to cause problems. Woven metal mesh is self-damping, so there is less concern for buzz and rattle. It is relatively simple to fabricate edge conditions for this material (see Figure 14.11).

Coiled Metal

I have used coiled metal, otherwise known as chainmail, for ceilings in a number of halls and had excellent results. This material is wire, so there is no issue with depth, buzz, nor rattle. The edge constructions are particularly easy to fabricate (see Figure 14.12).

Figure 14.11   Woven metal mesh, 51% open. Acoustically excellent, but may be too open visually. Can be backed with sound transparent fabric for a more opaque look.

Figure 14.12   Coiled metal mesh, approximately 66% open. An excellent choice for sound transparent materials.

Glass Fiber–Reinforced Gypsum/Concrete

These materials are problematic as the inherent depth of the material required for fabrication requires very open percentages. If off-axis attenuation is a concern, avoid these materials—even with large openings. Glass fiber–reinforced gypsum/concrete is acceptable over fixed acoustic absorption.

Fabrics

Thousands of sound-transparent fabrics are available, and a simple transmissibility test will determine attenuation at high frequencies. Avoid fabrics that are backed or so tightly woven that they will attenuate high frequencies. Blowing smoke through the material can be a dramatic test, but I find the test to be unreliable in critical applications (see Figure 14.13).

Testing the material for sound transmissibility for 20–20,000 Hz in an anechoic chamber is straightforward. Place a broadband sound on one side and a calibrated microphone on the other. This is less useful for large perforations and for thick materials. Non-incident attenuation is also difficult to measure accurately.

Figure 14.13 Smoke test for sound transparency. While smoke may pass through a material, this alone will not determine if it is sufficiently sound transparent.

Myths and Misconceptions

Myth #1: Unoccupied Seating Should Be Designed to Closely Match the Absorption of the Seats When Occupied

False. Seats should have minimal absorption in a multi-use hall. Adjustable absorption materials can be used to control the rehearsal mode acoustics.

Myth #2: If the Hall Sounds Just Right Empty or in Rehearsal Mode, It Will Sound Fine When Occupied by an Audience

False. A well-designed multi-use hall will always drop about 0.2–0.3 seconds of RT (mid frequency) with 80%–100% of the audience present. The hall should sound overly bright and reverberant when empty. This is normal. If it sounds perfect in rehearsal with all acoustic material stored, then it will be too dead when occupied.

Myth #3: Pony Walls, or Knee Walls, in the Seating Area Reduce the Negative Effect of Seat Dip Acoustic Absorption

False. The seat dip effect is a function of row-to-row spacing and the seat back height. Short pony walls have no effect. Actually, these low walls serve no real function except to reduce the space for the audience's feet, thereby increasing the required row-to-row depth.

Myth #4: Fabrics That Pass the Blow-Through Test Are Sound Transparent Enough to Be Used Anywhere

Partially true. Tests have shown that high-frequency attenuation can still occur even if breath or smoke passes freely through the material (see Figure 14.13).

MEASURING RESULTS

Chapter 15

Making It Multi-Use

Introduction

A hall becomes multi-use through the use of adjustable acoustic systems both on stage and in the hall. The physical systems described in this chapter control reverberation time (RT), reflection patterns, and loudness. Systems have tried to manipulate volume, but the best tools for the acoustician are absorptive treatments such as drapes, panels, and banners.

Manipulating Volume

RT can be reduced by decreasing volume; however, volume reduction systems are ineffective, costly, prone to failure, and potentially dangerous. Moving massive ceilings over audiences has been widely abandoned.

RT can be increased by adding acoustic chambers and volumes to the hall that are coupled to the hall through the use of massive doorways. In my experience, adding volume chambers via complex doors is not cost effective. Acoustic adjustments can be accomplished with better control, lower construction costs, and more predictable outcomes through the use of absorptive treatments such as drapes, banners, and panels.

Manipulating Absorption

Systems with movable or deployable sound absorption materials are true assets to the acoustician. Of course, in this case, hall volume and RT must be designed for the maximum value needed by symphonic events or choral concerts. This includes an RT of about 2.0 seconds mid frequency and a bass ratio of 1.2–1.4. Ideally, the volume in the hall will achieve these standards when all ducts, openings, carpets, and occupied seats are accounted for calculations, and adjustable acoustic materials are stored out of the hall volume. Adjustable acoustic systems must have enough surface area exposed when deployed to reduce the mid-frequency RT to about 1.2–1.4 seconds with an audience present.

ADJUSTABLE ACOUSTIC MATERIALS	63 Hz	125 Hz	250 Hz	500 Hz	1000 Hz	2000 Hz	4000 Hz
acouStac 3W11, 15 in. (375 mm) from wall	0.20	0.45	0.70	0.90	1.00	1.00	1.00
acouRoll RW12, 15 in. (375 mm) from wall	0.20	0.60	0.80	0.85	0.95	1.00	1.00
Velour, 25 oz., Single 100% gathered, 15 in. (375 mm) from wall	0.36	0.52	0.75	0.79	0.90	0.93	0.93
Velour, 25 oz., Single 100% gathered, 5 in. (125 mm) from wall	0.24	0.34	0.73	0.94	0.83	0.90	0.87
Wenger Transform, Velour 22 oz., 4 in. (101.6 mm) Air Space	0.25	0.36	0.79	0.90	1.03	1.05	1.07

Figure 15.1 Typical sound absorption coefficients for acoustic drapes and banners.

It might seem possible to target a greater variation in RT, especially for amplified events, by dropping it from 1.1 to 1.0 seconds. However, a 100% drop in RT involves deploying an unwieldy amount of soft acoustic materials as much on every wall surface in the hall and the ceiling. This is architecturally unfeasible.

Sound systems specifically designed for amplified performances in rooms with 1.2–1.4 seconds of RT can be highly effective and intelligible.

Low-Frequency Absorption

Utilizing adjustable acoustic systems to control low frequencies, particularly below 200 Hz, is challenging. The most efficient low-frequency absorber is double-layer acoustic banners made of wool serge placed approximately 6 in. (15 cm) apart and located about 15 in. (38 cm) off a wall. Alternatively, 4 in. (10 cm) thick glass fiber panels moving on tracks along the walls or in the reverberant volume are effective at absorbing more of the reverberant energy below 200 Hz. Straight line acoustic drapes crossing the center of the hall attic are efficient at absorbing mid and high frequencies but do not absorb the bottom octaves (see Figure 15.1).

Location

The location of adjustable acoustic systems within the hall volume is pertinent to the acoustic success of a hall. Placement depends on whether the ceiling is open or closed; the availability of catwalks or platforms for access; and the availability of zones unobstructed by building structure, building services, doors, or circulation paths. Optimal locations for acoustic drapes for the Annenberg Center are shown in Figures 15.2 and 15.3. This hall is described in the case studies section (Appendix XI).

Space for adjustable acoustics must be set aside early on in the design process. Drapes may impact seating plans, so allocate 18 in. (46 cm) around the inside of the hall on all levels for drape, banner, and panel travel. Ensure that the location of ducts, roof structures, and roof drains will not obstruct the systems. Discuss areas of potential conflict early on with the design team and with structural and mechanical engineers.

Proscenium Wall

The audience side of the proscenium wall is one of the best spots for acoustic drapes. In this location, drapes can be very tall and are made more efficient at low frequencies when located 15–18 in. (38–46 cm) off the wall. If this drape is bisected by rigging devices, such as the

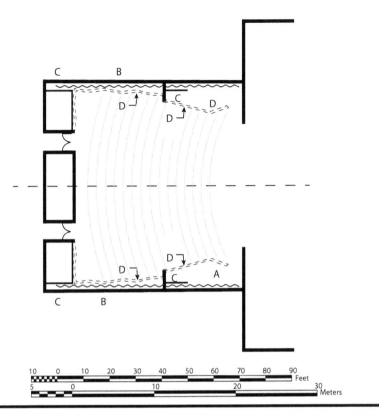

Figure 15.2 **The Wallis Annenberg CPA, Beverly Hills, CA, 2013. Preferred acoustic drape locations. (A) Drapes close to the proscenium on sidewalls. (B) Rear sidewalls. (C) Acoustic drape pocket. (D) Sound transparent wall system.**

forestage grid, then the drape can be split and located both below and above the grid. It is more efficient for drapes to be located in the large open zones in the attic of an open ceiling design rather than on small tracks that must fit around complex roof trusses. Fabric is far less expensive than motorized tracking. If drape tracks run parallel to large roof trusses, ensure that there are no structural ties or braces between the lower truss cords because these will obstruct the drapes. Acousticians have been fooled by an apparent wide open zone that is later found to be blocked by required steel members on the bottom cord running perpendicular to the trusses. Drape space might be available if the bottom chord member can be located only on the center line, and the drapes split left and right. Alternatively, use vertical moving acoustic banners that can easily avoid steel obstructions, but be aware that this adds significant equipment costs over the use of simple lateral drape tracks.

Rear Walls

Wall banners or drapes are less used on rear walls behind the seating except at the very top of the hall. The walls on lower levels are broken up with entry doors and control room windows that make it difficult to place treatments effectively. Energy returning from this wall, especially if it is

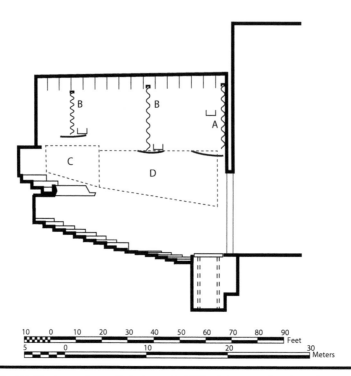

Figure 15.3 The Wallis Annenberg CPA, Beverly Hills, CA, 2013. Preferred acoustic drape locations. (A) Drapes on proscenium wall. (B) In reverberant upper volume. (C) Rear sidewalls. (D) Front sidewalls. All store in pockets.

under the balcony, and it has little result on the overall RT. If the rear wall is shaped properly, there is no echo to be controlled (see Figure 15.4).

Near Source of Speakers and Instruments

Materials that control loudness and harshness from speakers and instruments must be located near the source on the stage or on throat walls.

Upstage Wall of Orchestra Pit

In order to temper percussion and amplified instruments, 100% full, 24–26-oz. drapes should be placed at the upstage wall of the orchestra pit in 8 ft. (2.4 m) wide sewn sections. These panels should be walk along so that individual panels can be adjusted behind specific instruments. Pockets are not needed as drapes are rarely stored. For more information on orchestra pits, see Chapter 9.

Absorption Systems

The best tools to absorb sound energy are lightweight, storable, movable, and cost effective. They must also be fire resistant, long lasting, and easy to use. This limits materials to theatrical fabrics,

Figure 15.4 Acoustic drapes at rear of seating levels control sound reflections and have a modest effect on RT. Motorized track moves drapes into pockets.

lightweight acoustic panels, and banner materials. Acoustic bladder materials and other exotic absorbers are new to the industry but can be effective.

I am often asked if it is possible to fix the absorptive materials in place and move a solid reflective panel over them. My response is that the solid surface must have excellent bass reflections. In order to achieve this, the solid panel must have significant mass and stiffness, which demands complex hardware and powerful motors and may affect the structural frame of the building. A cost–benefit calculation favors utilization of carefully designed, lightweight materials.

Entry Doors

At the Smith Center in Las Vegas, Nevada, dozens of entry doors to the box tier's sound-lock vestibules are motorized and open during amplified productions. This reduces RT significantly and can be effective at reducing low frequencies. This is a clever option when used in conjunction with other systems and provided that the hall has adequate door space.

Horizontal Tracking Drapes

Straight track systems are significantly less expensive than tracks that curve alongside architecture. The track, motor, and control systems are the most expensive pieces of the assembly. Maximize the length of the cloth vertically for each drape track since textiles are less expensive than the track. Motors are usually half-horse power devices that move the drape horizontally back and forth.

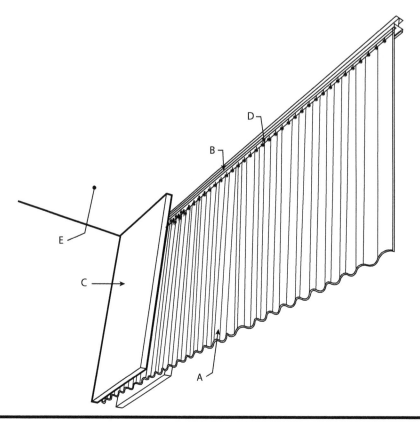

Figure 15.5 Typical acoustic drape system. (A) Motorized horizontal drape track system. (B) Structure above to support track. (C) Acoustic drape pocket. (D) Acoustic drape suspended from track system. (E) Wall 18 in. (0.5 m) off center of drape track.

These systems use simple, cost-effective controls that are easy for the theater consultant to design (see Figures 15.5 and 15.6).

Manual versus Motorized

Manual pull drapes present a lower-cost option. They are most viable in academic learning environments with inexpensive labor. Motorized drapes have far superior repeatability and consistency of settings and so are the recommended choice. Drapes hanging parallel to one another in free space should be spaced so that they are no closer than the vertical length of each drape, as shown in Figure 15.6. They can be stored in banner boxes to reduce the absorptive effects of the material.

Vertical Tracking Banners

Vertical tracking banners are more costly than horizontal ones because a more complex rigging and motor system is required. The cost is higher; however, these types of banners are more efficient at mid to low frequencies so that more absorption can be accomplished with less material. Banners typically pull up into a solid enclosure box so that they do not affect

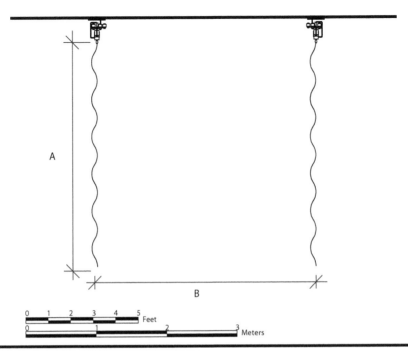

Figure 15.6 **Vertical acoustic drape spacing. Space acoustic drapes no closer together (B) than they are tall (A).**

the hall's RT. The acoustician must include absorption from the banner box in calculations. The fabric bunches as it retracts vertically, so a large opening is needed for banners to drop through a finished ceiling or soffit. Roller banners store more compactly than bunched banners (see Figure 15.7).

Roll Banners

Roll-down banners present another option for vertical solutions. They are more expensive than stacking drapes and consist of double-layer fabric spaced 4–6 in. (101–152 mm) apart. The manufacturer utilizes double rollers nested so that they take as little volume as possible. Roll banners store compactly and are most effective when the banner box is exposed in the hall. The case study on Baldwin Auditorium (Appendix II) goes into more depth on this type of banner (see Figures 15.8 and 15.9).

Banner Fabric

Banner fabrics include wool serge, velour, and roll-down sun shade fabrics. Wool serge is predominantly used since it is fireproof, can be custom dyed, and is dimensionally stable. Its acoustic performance is similar to stacking velour banners when wool serge is used in double layers spaced off the walls.

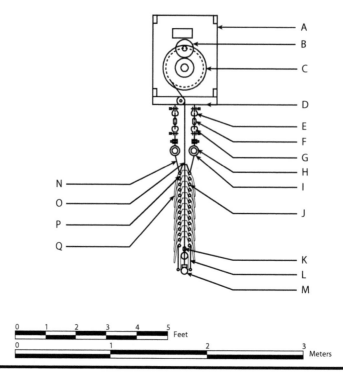

Figure 15.7 Dallas City Performance Hall, Dallas, TX, 2012. Alternate banner system. (A) Steel housing. (B) Drive motor. (C) Cable drum. (D) Double-layer wool serge. (E) Steel cable. (F) Steel batten at bottom drops through opening in ceiling. (G) Batten and wool serge banner can extend down 30 ft. (9.2 m).

Wool Serge

Wool serge materials are highly efficient (NRC 1.0) when stretched taut and used in double layers 4–6 in. (100–150 mm) and spaced 12–15 in. (300–400 mm) apart from a wall. This material commands the highest level of absorption possible, almost 100% absorption at all frequencies, and 80%–90% efficiency in the vital 100–200-Hz zone. Quality wool serge materials are long lasting and visually pleasing. The acouStac and acouRoll products are essential to the acoustic success of Alice Tully Hall at Lincoln Center in New York City. The Alice Tully Hall case study (Appendix I) covers this in detail.

The wool fabric is sewn into panels, is separated by metal spacers to hold air in the space, and uses retracting cables between the wool layers. Jaffe Holden helped develop and fine-tune this product in collaboration with the manufacturer in order to obtain maximum effectiveness. At Alice Tully Hall, the motors were upsized and synchronized so that banners would move at the highest possible speed and have the most exciting impact (see Figures 15.10 and 15.11). Standard motors move the banners at a more leisurely pace.

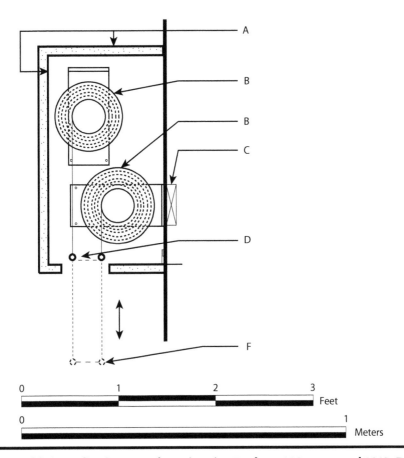

Figure 15.8 **Baldwin Auditorium at Duke University, Durham, NC, renovated 2013. Double roll acoustic banner. (A) Decorative wood enclosure. (B) Wool serge on motorized drum. (C) Secure mounting to wall structure. (D) Battens at bottom of roll. (F) Double-layer fabric drops over wall surfaces.**

Velour

The simplest and most inexpensive systems are velour drapes of 24–26-oz. material as used on theatrical stages. Used without lining, this textile is acoustically efficient (NRC 0.7) especially when located a short distance off a wall. It must be 100% full, which means that double the amount of material is needed on a track. Velour drapes are not particularly attractive and are often relegated to the attics of open ceiling systems or placed behind sound-transparent wall materials. However, the use of black fabric, tracking, and hardware can make them almost invisible. Performance is modestly improved when layers are doubled up, or lining is added and is not worth the added cost. Using heavier velour (30 oz.) is more expensive and has not proven to be more acoustically absorptive.

Figure 15.9   Double roll acoustic banners, partially deployed. Roller system allows for a more compact storage. The weight of pipes at bottom of wool serge holds fabric taught and maintains separation between layers and distance from wall.

Roll-Down Fabrics

If a more visually transparent look is desired, flow resistance fabrics offer a solution. MechoShade makes a product called AcoustiVeil. This material is less efficient than wool serge (NRC 0.6), and its absorption is less consistent across frequency bands. It rolls on standard window shade mechanisms, and although it is not very absorptive, it is cost effective.

Rigid Acoustical Panels

On occasion, glass fiber acoustic panels on tracks can provide absorption in attics of open ceiling spaces. These large panels are highly efficient except the lowest of frequencies. Rigid acoustical panels are stored in pockets in the center of the hall and moved across the attic on motorized horizontal tracks. This technique was successfully used in the Globe-News Center for the Performing Arts in Amarillo, Texas, in addition to the Benjamin and Marian Schuster Performing Arts Center in Dayton, Ohio (see Figures 15.12 and 15.13). At the Schuster, rigid panels were placed on three tracks on the throat walls. Panels were pulled out one at a time to cover the 12 ft. high (3 m) wall. The Schuster Center case study (Appendix III) discusses this in more detail.

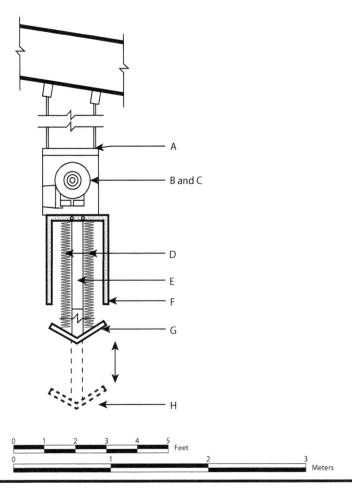

Figure 15.10 Wagner Noël PAC, Midland, TX, 2009. Alternate banner system. (A) Steel housing. (B) Drive motor. (C) Cable drum. (D) Double-layer wool serge. (E) Steel cable. (F) Wood banner enclosure. (G) Steel bottom closure. (H) Bottom and wool serge banner can extend down 30 ft. (9.2 m).

Banner Storage Boxes

Banner storage boxes can absorb significant amounts of low-frequency sound that the acoustician must account for in calculations. A hall with concrete or rigid masonry-backed drywall and a concrete ceiling can tolerate modest low-frequency absorption. In this case, it is possible to build ¾-in. (1.9-mm) plywood banner boxes in attic spaces.

Drape Pockets

Drape pockets are generally one-fourth the length of the drape track. This allows ample space for motors, pulleys, and other mechanisms as well as the drape itself. A double- or triple-layer drywall construction is necessary to prevent low-frequency absorption. Pockets made of masonry or concrete have little to no frequency impact.

Figure 15.11 Alice Tully Hall, New York, NY, 1969. Acoustic banners. High-speed double-layer wool serge banners drop through ceiling openings to control RT.

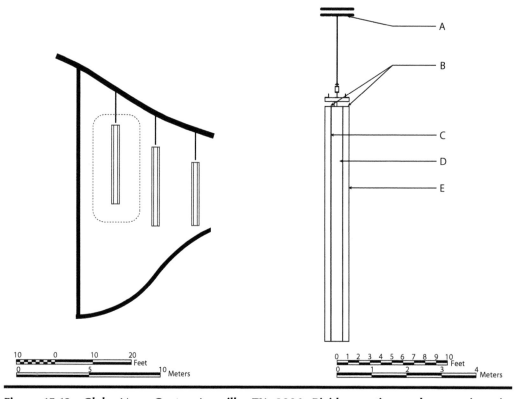

Figure 15.12 Globe-News Center, Amarillo, TX, 2006. Rigid acoustic panel system in attic. (A) Roof structure. (B) Track structure. (C) Drape pocket system in center of hall. (D) Framed acoustic panels.

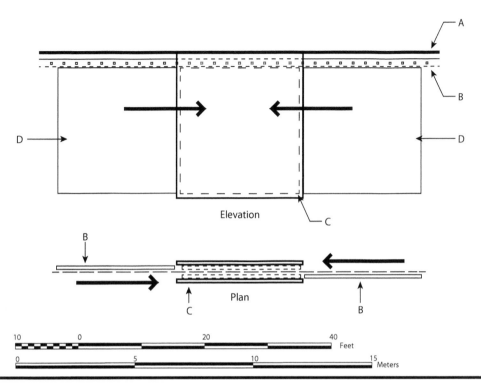

Figure 15.13 Globe-News Center, Amarillo, TX, 2006. Rigid acoustic panel system in attic. (A) Roof structure. (B) Track structure. (C) Drape pocket system in center of hall. (D) Framed acoustic panels.

It may seem counterintuitive to locate the drape pocket in the center of the longitudinal drape track. However, the center pocket can hold two side-by-side drapes in half the size and therefore half the level of pocket absorption. Place the pocket structure on the outside because the pocket will be smaller, and the drape will run smoothly over the interior surface without snagging on metal studs.

Pocket Doors

It is not necessary to include motorized pocket doors to close off the drape when it is stored. Pocket doors are an unnecessary expense since the absorption posed by the exposed end of the drape in the pocket is minimal. When requested by architects, I have used pocket doors for visual masking of stored wall drapes. They serve no acoustic purpose but offer a more pleasing appearance. The closed ceiling design of Bass Hall in Fort Worth, Texas includes pocket doors.

Motors and Control Systems

A motorized system is preferred for multi-use halls. Collaborate with the theater consultant to design a control system that allows for consistent drape adjustment from the stage or a control room. Cumbersome controls or unreliable systems are of little use to operators.

Control systems should have simple, easy-to-use graphic interfaces to help users select preset drape settings. The acoustician designs these settings during tuning and includes them in the

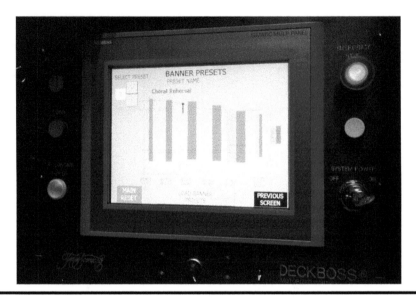

Figure 15.14 Baldwin Auditorium at Duke University, Durham, NC, renovated 2013. Controls for adjustable acoustic systems. Touch panel control allows users to select preset acoustic settings by performance type.

owner's manual. Rehearsal settings can simulate the effect of an audience in the hall so that sonic change between rehearsal and performance modes is more modest.

Settings

Control settings should have at least three settings per drape and four settings per banner. These include stored, deployed, preset 1, and preset 2 (for banners only). Ganging together of banners and drapes is recommended for simplicity in tuning and operation. Systems are labeled stored or deployed to avoid confusing terms like in, out or up, down (see Figure 15.14).

Drapes and banners are excellent devices for controlling RT and G levels as well as reflections; however, small changes in the acoustics are often unperceivable. Although dozens of presets are possible, only about six settings can be perceived by a sophisticated listener.

Myths and Misconceptions

Myth #1: Small Areas of Adjustable Acoustic Materials Can Make a Big Difference in Acoustics

False. To make significant changes to RT in a multi-use hall, thousands of square feet or hundreds of square meters of materials must move in and out of storage. Small areas do not make a big difference.

Myth #2: The Exact Location of the Materials Is Not Important; Just Add Them Wherever Possible

False. The location of adjustable acoustic materials strongly affects their efficiency and performance. In order of decreasing effectiveness are upper side walls, proscenium wall, lower side walls, crossing of the attic zone, and finally, rear wall.

Myth #3: Increasing the Weight of the Drape or Banner Dramatically Improves Acoustic Absorption

False; 24–26-oz. velour works well. Increasing the weight to 30 oz. or more has a big impact on cost but minimal impact on acoustic improvement.

Myth #4: The Greater the Spacing of Materials Off a Wall, the Better the Low-Frequency Performance

Not true. Data suggest that 15–18-in. (380–460-mm) spacing off the wall maximizes absorption. Increasing the air space beyond this will reduce performance.

Myth #5: Acoustic Banners Provide Limited Modification of Acoustic Performance Compared to Reverberant Chambers

Wrong. Figure 15.15 depicts field measurements recording very large changes in RT with banners at Dallas City Performance Hall in Texas.

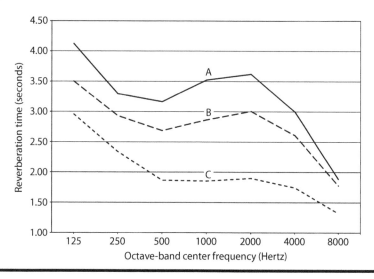

Figure 15.15 Dallas City Performance Hall, Dallas, TX, 2012. Reverberation time measurements. (A) Banners stored, shell in. (B) Half banners, shell in. (C) Banners deployed, no shell.

Chapter 16

Electronic Architecture Systems

Introduction

Electronic architecture (EA) systems, also known as electronic enhancement systems, are a viable substitute for the adjustable physical acoustics methods covered in Chapter 15. EA systems may eventually replace the practice of using large banners, drapes, and panels if they continue to evolve and gain acceptance. These systems are not without controversy and detractors especially since they are often unacceptable to implement in a new multi-use hall design. EA system techniques are much more common in renovations where physical approaches would be impractical or unaffordable.

My firm and Chris Jaffe were early pioneers in the design and use of such systems in the 1970s electronic reflected energy systems (ERES) in order to provide reverberation time (RT) adjustability in lieu of moving drapes and banners. The technical systems have greatly improved in quality and reliability, but the negative attitude toward EA systems from purists, music critics, and musicians has been slow to change.

Technologically Sound

From a technological standpoint, EA systems make sense. They use high-quality microphones, very sophisticated signal processing, and multiple loudspeakers to deliver enhanced RT, C80, envelopment, and bass ratio (BR) at the touch of a button. Rather than moving vast amounts of soft goods with motors to reduce RT, the EA system establishes a lower RT base line and adds RT and other subjective factors electronically as necessary for the type of performance. It is logical and more economical to build a more compact hall with its lower V and less massive interior surfaces in order to achieve a lower base line RT as well as BR. There may be financial savings in cutting physical acoustic systems, power, controls, and access catwalks and replacing them with electronics. Maintenance of motors and tracks and the periodic removal of drapes for refireproofing are eliminated. Yet, EA systems are slow to be adopted in new halls.

Early Use

When EA systems were first rolled out, the technology was unreliable despite being advanced for its time. Early EA configurations including the Parkin Acoustic System used Helmholtz resonators at the microphones in an attempt to dampen feedback. Ultimately, coloration and odd ringing-on notes occurred when this system was installed in the ceiling of the Royal Festival Hall in London, England. The systems sounded just natural enough to get by, but slight changes in the location of a resonator tube, or even a shift in temperature or humidity, could add audible coloration. Music critics had a negative response to the system.

The first EA system in the United States was designed by Jaffe Holden in the early 1970s for a historic movie palace in Oakland, California. This system used early reflection speakers over the first shell ceiling and mid- and low-frequency reverberation speakers behind decorative ceiling grilles. This temporary installation proved the effectiveness of the EA system for halls with low RT and poor C80 reflections for symphony orchestras (see Figure 16.1).

Reaction from Musicians

Some musicians felt that EA systems altered their sound in ways that were not natural and not under their control. Some reacted distastefully to the notion that electronics were involved regardless of the quality of the sound produced.

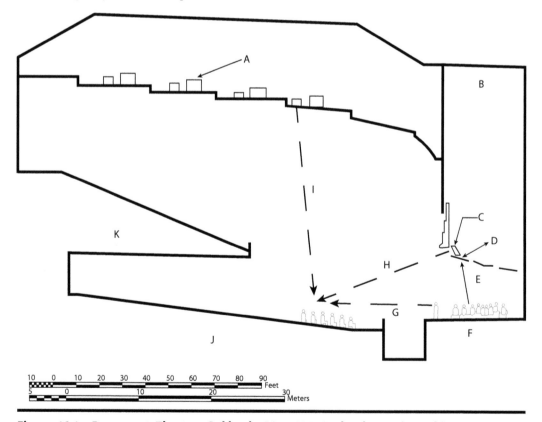

Figure 16.1 Paramount Theatre, Oakland, CA, 1931. Early electronic architecture system from the 1970s. (A) Microphones in two locations in the shell. (B) Ceiling reverberation speakers. (C) Underbalcony speakers. (D) Lateral energy speakers for C80 reflections.

Confusion swirled around two types of systems: EA systems and sound reinforcement systems. Both used similar microphones, processing, amplifiers, and loudspeakers. Musicians feared that they might lose artistic control if an operator amplified or modified their sound and wrongly attributed this fear to EA systems. In fact, EA systems have no such operators.

EA systems provide acoustic reflections that would not otherwise be provided because of the hall's limited volume, surfaces, shape, or audience orientation. Reflections and reverberation that could be provided by a taller ceiling, for example, can be simulated by speakers placed in the lower ceiling volume. The direct sound from the performer to the listener is never compromised; the reflections added would normally occur from room surfaces.

Early systems were remarkably successful in augmenting RT and adding early reflections for improved clarity. Each generation had less coloration, was more stable, and had better acoustic performance controls. Ongoing service, retuning, and reliability were often the downfall of early systems. Technicians could tweak and misadjust the circuits or the controls.

Prejudice toward these systems remains today even as vast improvements have been made in reliability, control, and naturalness. I now see modest acceptance from the musical community regarding renovations of acoustically deficient halls and in musicians' practice rooms. I believe that over time, these prejudices will fade as more progressive-thinking artists and managers start utilizing electronics in new ways (see Figure 16.2).

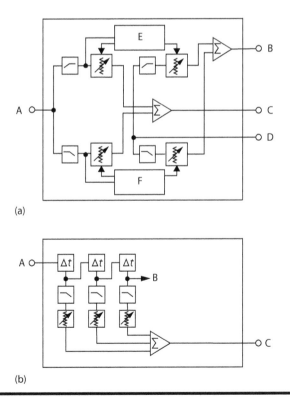

(a)

(b)

Figure 16.2 **(a) RODS circuit and (b) 72 tap delay. This was used in the 1990s to create and control reverberation. (A) Signal from mic pre-amps. RODS had a LF (E) and HF (F) section with a feedback loop to control levels. The 72-tap delay was made by Industrial Research and created delayed replicas mimicking reverberation.**

EA System Components

Certain equipment in all EA systems can be obtained from a number of excellent manufacturers or, if desired, custom designed by an acoustician using off-the-shelf components. The EA system is one of many tools in an acoustician's tool box and should be selected based on factors such as cost, function, reliability, service, and performance (see Figure 16.3).

Signal Processing Unit

At the heart of the EA system is the signal processing unit. It adds natural-sounding RT without coloration or feedback. All systems utilize distributed microphones and speakers throughout the performance stage and hall. By definition, microphones pick up audio from speakers as well as performers and negative recirculation. This can cause coloration and horrible feedback or a howl. The signal processing unit must control this unwanted coloration and aid in gain before feedback. It must also add stability to the system and provide levels of adjustment on RT, C80, envelopment, BR, and other factors.

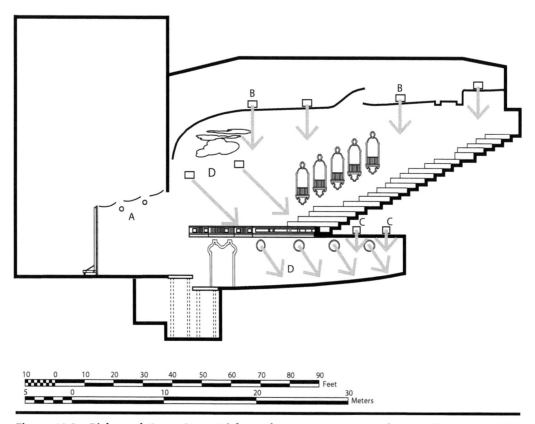

Figure 16.3 Richmond CenterStage, Richmond, VA, 1928, renovated 2009. Electronic Architecture (EA) device locations. (A) Shell microphones on winches. (B) Ceiling reverberation speakers flush mounted. (C) Underbalcony flush mounted speakers in dense pattern. (D) Early reflection speakers hidden inside walls.

Control System

Detailed descriptions of EA control systems on the market are beyond the scope of this chapter. Many of these systems are confidential and proprietary. All share in these common functions:

- Added reverberation without coloration or resonances
- Freedom from feedback, howling, and spurious noise
- Limited user controls but full controllability for tuning
- Numerous presets for varying RT, C80, envelopment, and BR
- Self-diagnostics and trouble reporting

Microphones and Loudspeakers

The placement, selection, and service of microphones and loudspeakers are critical to the success of the EA system. Each proprietary system has its own standards for quantity of microphones and their locations. Some use a larger quantity of mics distributed around the stage and near the ceiling and walls of the hall. Others place mics close to the stage alone. One system uses a quad array of microphones (four mics packed closely together; left, right, up, and down) placed in several locations.

The placement of loudspeakers in a multi-use hall is always a challenge. Integrating these units in the renovation of a historic theater presents an even bigger challenge. My solutions for this include decorative painting of units, recessing them in ceilings and walls, and even locating them under seats in air plenums in an effort to make them invisible to the eye yet fully audible (see Figure 16.4).

Figure 16.4   Richmond CenterStage, Richmond, VA, 1928, renovated 2009. Early reflection speaker. Locating early field speakers in decorative walls is challenging. Here a small speaker is hidden behind a decorative bust but precisely aimed.

Using EA Systems Today

Jaffe Holden has designed a number of multi-use halls that successfully utilize the EA system's unique capabilities to adjust RT and other acoustic parameters. Examples range in size and include Richmond CenterStage's Carpenter Theatre in Richmond, Virginia (see the case study in Appendix IX for more information), the 22,000-seat Assembly Hall for the Church of Latter-Day Saints in Salt Lake City, Utah, and a 220-seat set for a popular late night show in Rockefeller Center in New York City. The unusual feature of these halls is the relaxation of the *V* criteria and the massive surfaces that sustain the BR. It is important to note that an EA system cannot adjust sound that originates from musicians and singers, so it cannot compensate for a poorly designed stage enclosure or orchestra pit (see Figure 16.5).

Figure 16.5 Richmond CenterStage, Richmond, VA, 1928, renovated 2009. EA equipment rack houses electronic components. Processing circuits are located on top, and amplifiers are in the lower section.

A hall's fundamental acoustic characteristics should follow the design criteria set forth here, as the EA system cannot overcome acoustic problems including excessive RT, low-frequency buildup, echoes, focusing, and excessive HVAC noise or poor sound isolation from exterior noise sources. In fact, the EA system may exacerbate HVAC noise picked up by microphones near loud grilles.

Guidelines for Designing Halls with EA Systems

- *Limit the V for RT in the 1.3–1.5 mid-frequency range for amplified and speech events with the EA system off.* This implies a lower ceiling and perhaps construction cost savings.
- *Avoid an excessive acoustic V that requires the addition of permanent absorption to reduce RT to these levels.* Absorption will reduce G and RT levels as well as BR.
- *EA systems relax criteria for BR to 1.0, thus allowing modest low-frequency panel absorption for walls and ceiling.* Thinner wood paneling or one less layer of gypsum board can then be used, resulting in lowered construction cost.
- *The underbalcony criteria can often be relaxed as two to three more rows of seating might be placed under balcony overhangs.* Underbalcony EA speakers can be employed to offset the choking effect of the low ceiling.
- *Speakers should be located in the hall ceiling, upper side walls, underbalcony areas, and sometimes under the floor in air plenums.* This produces an even, natural-sounding RT. Generally, audiences should be in the field of at least three EA speakers. In underbalcony areas, this can add up to a large number of units. Early reflection speakers require careful placement to enhance C80 and lateral energy.
- *EA systems require a commodious rack room with significant power and cooling.* This room should be located in close proximity to the hall in order to minimize cable routing.

Settings

Once installed, these systems provide acoustical enhancement for various program functions. For example, when the natural acoustics of a room are very dry, there could be a setting for speech, for highly reinforced music presentation or concerts, for an opera that supports singers and musicians in the pit, and for orchestral and choral performances. There might also be variations of each of these settings. It is very important for the acoustician to develop the settings and parameters that will best support the program functions over time.

Tuning

Proper tuning and adjustments are critical to the success and usability of the system. Jaffe Holden's AV designers have extensive experience tuning these systems. They found that since EA systems are made from the same hardware as audio systems, audio operators assume that they know how to tune them. It is my experience that very few audio operators know what a great natural hall is supposed to sound like, and even many acousticians only have a theoretical paradigm of hall sound to draw upon.

The brain, the ears, and the eyes must all agree that what we hear makes sense. In theory, it may work to drop an RT curve into a hall, but in reality, that is not enough. When acousticians tune a hall, we start with what the manufacturer provided, but it is our job to take it to the next level. It takes the ear of an experienced acoustician with the reference framework of truly great halls to polish a room with an EA system so that the sound is technically correct and holistically appropriate. A good room does not just sound good; it also sounds at one with itself.

Myths and Misconceptions

Myth #1: EA Systems Are Good but Never Really as Good as Natural Acoustics

This may have been true of systems in the 1970s and 80s, but modern systems are indistinguishable from natural acoustics. EA systems cannot make a terrible hall good, but they can make a decent hall that lacks RT or proper reflection patterns exemplary.

Myth #2: EA Systems Are Unreliable, and When They Fail, There Is No Way to Fix Them

Early systems did have reliability issues. Modern systems are very well supported by their manufacturers, and malfunctions are now rare. Malfunctions can be diagnosed remotely and serviced by a local representative or contractor. Extended warranties up to five years can be purchased.

Myth #3: Musicians Will Never Accept EA Systems

Since 2013, I have noticed greater acceptance from musicians and management. Perhaps younger musicians are more accustomed to electronics and less afraid of combining music with technology. I have seen acceptance and even enthusiasm in cases where use of an EA system is mandated.

Myth #4: Systems Are Too Expensive

As competition increases between manufacturers such as Meyer, Wenger, ACS, and E-coustic, prices have been dropping. Control systems can utilize common tablet devices to lower costs as well. Prices for these types of EA systems are comparable or slightly higher than adjustable banners, drapes, or panels. Due to lower volume and lighter-weight constructions, these systems look very attractive and have added savings.

Myth #5: EA Systems Will Never Replace the Orchestra Shell on Stage for Classical Musicians

The electronic shell is used in outdoor locations quite successfully. It provides excellent onstage hearing, blending, and ensemble reflections when the stage area is nonreflective or poorly shaped. Indoor electronic orchestra shells will soon follow.

Chapter 17

Tuning the Hall

Introduction

Adjustable acoustic systems in a multi-use hall are most effective when they are set at the appropriate position for a particular program. Tuning the hall is the process of determining the optimal position for the musicians on the stage, as well as the shell, banners, and drapes for each type of performance or rehearsal. Calculations and modeling can determine rough, general settings, and measurements can then prove the impact on reverberation time (RT), early reflections, and other acoustic criteria. Tuning the hall to its optimal settings requires the use of live musicians both onstage and in the orchestra pit.

Collaborative Process

The process of tuning is one that has developed over many years through evidence-based results in dozens of facilities. No two halls are identical. Every hall has unique systems and attributes, so it is not possible to pinpoint one methodology that achieves optimal sound in all halls. That process itself is collaborative and team based, much like the initial design of the hall.

Questions must be asked. The acoustician must listen to opinions and integrate the observations of musicians, experienced listeners, and other team members. During the tuning of Alice Tully Hall at Lincoln Center, my team gathered input through the distribution of written questionnaires to faculty, board members, and experienced listeners. At other halls, we were less formal and asked musicians specific questions about shell settings, musician positioning, and the way the hall responded to drape settings.

I believe that tuning a hall is similar to tuning a piano. The piano tuner begins by forcing the string out of tune and then slowly brings it into tune.

The basic approach begins with the acoustic banners fully extended in the hall. Gradually, the banners move to a position with an improved condition. Start with the instruments fully upstage, and gradually move them downstage into more optimal positions so as to hear the differences and understand how the hall is working. Interestingly, the same methodology of tuning is used for tuning electronic enhancement systems. With electronics, the RT is set for unnatural longer and louder reflections, and then they are backed down until the sound is natural and comfortable.

Preparing for the Tuning

Many technicians and musicians have been gathered for the tuning, and the acoustician must be prepared.

What to Know

Be familiar with all calculations and measurements before arriving. This includes checking ceiling reflector angles, ray tracing or CATT models, RT measurements, and background noise measurements. Study the drawings for acoustic drapes and banners. Know how they are labeled, where they are located, and how they are controlled. Have a thorough understanding of the control systems. Know where they are located and how they work.

What to Bring

Do not assume that drawings will be provided by the facility. Bring half-size drawings of key items. A flashlight and a 50-ft. tape measure will be invaluable in addition to tuning tools including the angle meter, meters, and acoustic balloons. Do not forget to bring a camera to document the process.

What to Do

Before the tuning event begins, speak with the technical director or manager to get an accurate feel for what works and what does not and what issues or problems have occurred in the facility. Create a tuning schedule, and review it with the technical director or music dean. Confirm when groups will be in the hall and what accommodations they will require. Try to get as much community involvement as possible. Utilizing local high school bands or community chorus groups during tuning generates excitement about the new facility. Gently discourage groups that will not be helpful in the tuning process. Schedule acoustic measurements as well as breaks.

When you arrive for tuning, set up a tech station at the center of the hall with power outlets, chargers, laptop, drawings, and chairs to create a command center. Make sure that a stage crew is handy, and be aware of their onsite schedule and union requirements for breaks.

Setting the Shell for Tuning

Setting the orchestra shell is the most time-consuming and complex part of the process. The shell has many functions; it must work acoustically, be visually attractive, mask backstage and offstage areas, allow access to musicians, allow for airflow, etc. The first step is for the theater consultants to check the shell to ensure that it is hung on the correct line sets, that the angles of the ceilings are close to our recommendations, and that lights in the ceilings are set and working correctly.

Next, bring the ceilings in to the stage level, one at a time, for visual inspection. The acoustician must check for warping, delaminating, damage, and sighting down the leading and trailing edge to see that the edges are straight and true.

Ceiling Angles

It is important to assess the angle of the reflector face as it relates to the stage floor. This is easily accomplished when the ceilings are lowered to about 4 ft. (1.2 m) off the floor. Check the difference in height off the stage floor for the leading and trailing edge of the shell (upstage/downstage) from that which is on the drawings. Trust the angle that ray tracing and CATT models suggest even if it looks wrong.

Typically, the angle of the large (downstage) part of the face of all three reflectors is set at 15° to horizontal as a starting point. The rear part, or the back third of the reflector, is the curved swoop that sends energy back to the musicians onstage, and the front two-thirds of the reflector sends energy out to the hall. We investigate the first two-thirds because the rear is not angle specific.

Each ceiling piece is a three-way valve, in a way, that operates in the time and energy domain. Sound is blended onstage to become homogeneous, directed out to the audience, and vented to the upper volume of the stage house in order to reduce the loudness of brass and percussion.

The ceiling piece located furthest upstage is the choral reflector, since the orchestra usually plays forward on the lift. Here, the angle setting is steeper—up to 20° or 25°—so that choral sound is projected into the audience chamber and does not get overpowered by the orchestra.

Next, determine the height of the ceiling reflectors from the floor. Using the calculations as a guide, take a 50-ft. tape measure to the leading edge of the shell and fly it out to the calculated height to determine the starting point. The ceiling reflectors must pass the visual and listening tests. When lights are in the shell ceilings, the lighting variable is removed from the equation. If lights are centered between the ceilings, the process becomes more complex as lights will need to be masked, angled, and set.

Forestage Reflectors

In many halls, the shell starts with the forestage reflector or the eyebrow piece. This extension of the orchestra shell ceiling has a slightly different function from that of the onstage reflectors. Forestage reflectors are located forward of the proscenium, and the volume above is designed to be part of the overall acoustic volume of the hall. The sound that travels through the reflectors is not lost in the stage house. It contributes later in time to the overall acoustic energy, or loudness, and drives the upper reverberant volume of the hall. If the forestage reflector is too tight, the upper volume is starved of sufficient sound energy and cannot create the proper level of reverberation. This will result in too much sound directed down to the audience and musicians, strings that are far too bright and harsh, and weak reverberation that lacks envelopment. Refer to Chapter 10 on shell design. Often, forestage musicians have a hard time hearing themselves and each other if the forestage reflector is too open, and there is already a strong reverberant field and great envelopment.

The first of the two reflectors (the one closest to the proscenium) should be set at about 20°–25° and the second reflector set at 15°. Utilize ray tracing to determine these positions. Sometimes, a winch is in place to provide vertical movement for the forestage reflector. This allows the array to move up and out of the way of lighting angles. If this is the case, the trailing edge of the reflector should start 30 ft. (9.1 m) off the stage. At times, the forestage reflector is fixed in position and the winch eliminated to save money on a project. This was the case at Dell Hall at the Long Center for Performing Arts in Austin, Texas. Here, we set the fixed forestage ceiling reflector at a higher position than optimal, about 35–40 ft. (10.7–12.2 m) off the stage. This works, but a movable forestage reflector on a winch is preferred. Note that the angle of the forestage can be preset from calculations, so angle adjustability is not normally required (see Figure 17.1).

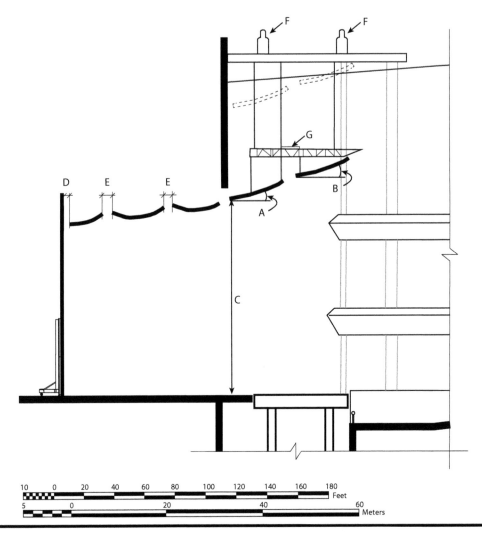

Figure 17.1 Wagner Noël PAC, Midland, TX, 2009. Movable forestage settings. (A) Set angles at about 20°–25° for 1st unit. (B) 15°–20°. (C) 30 ft. (9.1 m) above stage level. (D) Maximum 24 in. (0.6 m) gap between shell and towers. (E) 18 in. (0.4 m) target gap between ceilings. (F) Rigging motors on forestage grid. (G) Steel rigging frame to support forestage reflectors.

Shell Tower Settings

Orchestra shell towers, in conjunction with shell ceilings, support the orchestra and chorus with strong reflections that improve hearing, support blending and reverberation, and project sound into the hall. Their acoustic, aesthetic, and practical design is discussed in detail in Chapter 10. Here, we put them into action.

Set towers according to the theater consultant's design drawings, but do not permanently mark them on the floor yet. The tower closest to the proscenium is tricky because too large a gap allows the audience to see into the stage house. Instead of dropping soft goods into that gap for masking, adjust the angles of the wall to be slightly more off stage so that the line of sight is blocked adequately.

As with every adjustment, moving the towers too far off stage affects other criteria. The gap between the towers and ceilings should not exceed 24 in. (0.6 m), or else the positive wall/ceiling reflection pattern will be negatively affected. Visual appeal is also lost if the gap between the walls and the ceiling is uneven or too large. The 24-in. (0.6-m) gap allows the towers to come off stage at the fire curtain and looks good from locations in the audience chamber.

Tuning Adjustable Acoustic Drapes and Banners

As with all halls, the reverberation requirements of the symphonic orchestra are ill-disposed to amplified music and Broadway productions. For the most part, a hall should be at the most reverberant with an RT of at least 2.0 seconds for classical symphony and choral music. A question presents itself—why not leave all the drapes stored and go with maximum RT for symphonic tuning? There are three factors involved in this answer:

1. *It is best to start with all drapes deployed to see how the room reacts and how the energy from the stage fills the room.* This allows the acoustician to hear the direct sound more clearly with less cover from the reverberant field.
2. *Leaving some drapes deployed provides a vital simulation of the audience condition during rehearsals.* This allows the acoustician to learn how much drape to deploy to accomplish the best sound.
3. *The upper balcony seats can be a bit too reverberant.* The deployment of one-to-three upper rear drapes might be necessary even in concert mode with audience.

Motor Controls

The advancement of software programming technology for motor controls leads to a temptation to produce an unlimited number of tuning options. Avoid this mistake. Primary end users have limited time and experience setting the system, and there is a big risk of using incorrect settings. Select controls that are designed to be intuitive to users and the tech crew and that list the type of programming being performed in the hall rather than using complex and confusing presets.

Examples of Presets

- *Symphony setting*: All drapes and banners are stored except for the upper rear drape at the rear of the balcony.
- *Opera setting*: Some ceiling banners are deployed.
- *Amplified music/voice setting*: All drapes and banners are deployed.

As much as we wish for fine-tuning control of individual acoustic criteria such as RT and early reflections, remember that acoustics are not controlled that acutely with any device. Acoustic banners work well, but they are a rather blunt instrument. At most, there should be two presets in between the stored and deployed setting. Many settings can be grouped together rather than individually controlled. Overly complex control systems are expensive, provide no real value to end users, and may not even be used.

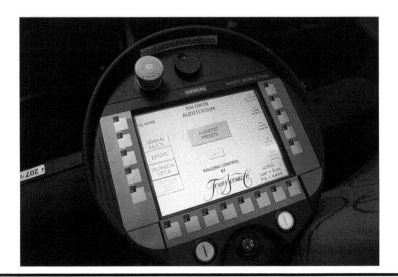

Figure 17.2   Image of banner control pendant that allows safe monitoring of moving acoustic systems.

Essential Features of a Control System (see Figure 17.2)

- *Control pendant*: This allows drapes to be controlled by a technician standing center stage and watching the drapes move. A duplicate control system exists in the control room.
- *Touch screens and presets*: These must be based on intuitive thinking rather than machine language.
- *Visual guide*: Since drapes and banners are not always visible to the technician, it is helpful to have a numerical readout of the percentage of movement for drape positions.

Rehearsal Mode versus Performance Mode

Since tuning of the hall occurs during rehearsal, consideration must be given to the different conditions surrounding a performance. You will need to account for this discrepancy when you complete the settings for the hall.

Plan to set the drapes so that sound is over the top, that is, too reverberant, too big, and too lush. When the audience enters, the hall will settle down and come right in to line. Rehearsal mode should be 0.2–0.3 seconds more reverberant than optimal to accommodate the impact of an audience. An audience also reduces direct energy and reflections. If you tune the hall during rehearsal mode for optimal sound without an audience, it will be low in reverberation and lack brightness and loudness in performance mode.

Introducing Musicians

I always find it thrilling to witness the first musical performance in a newly designed facility. Remember that tuning a hall is like tuning a piano—the piano tuner begins by forcing the string out of tune and then brings it gradually into tune.

Following that basic idea, begin with all drapes fully deployed. Understand that the hall will never be used for unamplified musical performance in this configuration as it will sound dead and muffled. Gradually store each group of drapes, listening to differences in sound in all parts of the hall. Listen not only for increased reverberation but also for the timbre differences, the sense of surround, the balance of low-to-high registers, the loudness, and the energy levels. Document the settings with acoustic instrumentation, and take images to record the process for later use.

With the shell in its full configuration, meaning with all ceilings and towers in place, start with musicians far upstage within the shell. Experience indicates that this is not best sound, but it is the starting point of a process that is aurally complex and time limited. The simple yet effective practice is to begin at the poorest setting and make incremental improvements.

Tuning with a Piano

A new hall often gets a new 9-ft. Steinway B or D model piano. The solo grand piano is ideal for tuning because it is percussive, full frequency, and dynamic. It moves easily about the stage and has a large repertoire of music to test, and a high-quality pianist is not difficult to find.

Finding the Ideal Location in the Shell

The forestage lift is often the best musician location because the hall is usually energized, meaning that the reverberation level is closer to the direct sound. Overall, the piano is louder, and better feedback to musicians exists. More shell support and better onstage hearing exist for musicians in an upstage position, but there is often a disconcerting lack of feedback from the hall itself.

Settings for Piano

Starting with all drapes deployed, move from bad to good. Tonality of the piano may change dramatically as the drapes are stored. Normally, the drapes should be set so that the sound is a little over the top, meaning a bit muddy in the lower register and lacking definition in the upper notes. The audience will settle the sound down to just the right level.

Settings for Choral Ensemble (12–15 Voices) with Piano

Start with a simple riser system, and position the ensemble in their normal performance mode but against the upstage wall of the shell. Musicians seem to like the way their sound is blended in this position, but remember, conditions are different outside of rehearsal. Move them downstage in increments until an ideal location is found. The location is ideal when the chorus sounds bright and clear, when vocal diction is well defined, and with a strong resonant component. A word of caution: a location too close to the audience may lose the vibrant support of the orchestra shell for the musicians.

Drapes should be settled into position after listening and consulting with the choral director and accompanist. Remember to set drapes at positions that may be slightly too reverberant with the anticipation that the hall will dry out with an audience (see Figure 17.3).

Figure 17.3 Annenberg Center for Performing Arts, Los Angeles, CA, 2013. Author listening to USC chamber orchestra for setting of ceiling reflector angles.

Symphonic Orchestra

The ceiling angles of the shell should be fine-tuned based on extensive listening onstage with the musicians. Flattening the ceiling reflectors will improve onstage hearing but comes at the expense of sound projection to the audience. Overly loud brass and percussion can be tempered by flattening the ceilings or by raising the rearmost ceiling element, if there is one, to bleed more sound into the stage.

Lowering ceiling pieces not only often improves onstage hearing for musicians but also results in higher onstage volume levels that throw sound off balance. When the ceiling is too low, sound becomes harsh and brittle in loud passages.

Once again, begin with the musicians fully upstage, and move them downstage in increments after listening to the rehearsal for 30 minutes or so. Musicians often prefer their own sound when in the far upstage position, but in many halls, this location lacks impact and vibrancy in the audience chamber and is visually unacceptable. Musicians voice the opinion that they want to be as close to the audience as possible, out on the lifts at the edge of the stage. This may ultimately become the preferred location, but it frequently presents issues with onstage hearing. Moving upstage about 10 ft. (3.0 m) from the stage edge often enables a beneficial early reflection off the stage floor. This is often the preferred performance location.

Acoustic banners and drapes are most likely in the fully stored position for symphonic performances but not always. I have found that leaving a rear balcony drape deployed is wise if the hall rarely fills to capacity during symphonic performances. Listening to rehearsals with the adjustable acoustic systems deployed is a diagnostic tool for hearing echoes, sound focusing off rear walls, early reflections, and distribution of direct energy that is indistinct when the hall is fully reverberant. Dallas City Performance Hall has side wall acoustic banners partially deployed for chamber music concerts.

For rehearsals, a partially deployed drape and banner combination better approximates the audience absorption effect. This means that the change when the audience arrives is less drastic.

Symphonic Risers

I often take a nontraditional approach to orchestral risers. Our shell ceiling designs make the risers almost unnecessary for onstage hearing. The last few stands of the violins can be raised if they cannot be heard well, and woodwinds and French horns can be placed on risers to help them hear and be more audible in the hall.

It is counterproductive to place brass and percussion on risers because they are the most powerful instruments in the orchestra in terms of raw sound power level (loudness), and they can easily overpower the strings. Why exacerbate balance issues by placing brass and percussion on risers?

Cello boxes, or resonant risers for each cello, can be useful when there is an apparent lack of cello energy in the hall. Bass risers can also add a few decibels to increase bass loudness in the hall.

Maestro Neal Gittleman, conductor of the Dayton Philharmonic at the Schuster Center in Dayton, Ohio, confided that he had tried every possible combination of orchestral risers over many years to find the best sound. Ultimately, he reverted back to our tuned setting of all musicians flat on the floor.

Risers are called for when there is a demand for better visual impact of the orchestra. Modern audiences want to see the musicians clearly, and risers facilitate this. However, risers must be accounted for in the tuning process because they will affect shell ceiling positioning.

Collaboration

A note on collaboration in tuning: I find working with artists and musicians to be thrilling and enlightening on a number of levels. This group listens in a way that is different from how engineers and consultants listen. They have vast experience listening to the direct sound of music but not necessarily to the hall alone or to its reverberation and reflections. Instead of open-ended questions like, What do you think?, I prefer to ask if they could hear themselves well and what differences they noticed between different locations. After listening intently and considering their feedback, I often find that the musicians were right.

The following is an excerpt from the author's journal describing his experience tuning the Wagner Noël Performing Arts Center.

Tuning; Wagner Noël Performing Arts Center at the University of Texas of the Permian Basin

I arrived in Midland, Texas with anticipation and regret. I was excited to finally hear the acoustics and work with live musicians and artists performing rather than just balloon pops and hand claps. I regret knowing that this may be the last time I am here and because it is so remote that few will visit it and appreciate what is here (see Figure 17.4).

The entire design team, contractor, and owners worked together in a partnership that allowed me, a kid from Hudson, Ohio, to look really good. Without great architects, builders, and owners, this would never have happened, and this great building that will stand for 50–100 years would not exist, and I would not have had a chance to participate. I get all the credit, but I alone really can do little. Now, millions of audience members will have a chance to be transported, to be moved and escape or be thrilled, or to just enjoy something very special. I am an incredibly lucky guy.

The hall is a cleanly designed acoustic diagram; it is simple and straightforward. The shell, lift, forestage, and drapes are tunable, whereas the rest of the walls, floors, ceilings, and balconies in the room—the vessel—are not tunable but were made so that these elements mesh perfectly with our adjustable systems.

This hall is stripped down to just what is needed acoustically but no less. Losing any single acoustic element would be a tipping point, and the acoustics would fail. The elements that are here, such as the wall shaping, are

Figure 17.4 Wagner Noël PAC, Midland, TX, 2009. Duet of clarinet and piano play in various stage locations to find the acoustic sweet spot.

optimized to best response but no more than is needed. The wall elements are shaped like oversized bricks and modulate in depth and size to provide mid- and high-frequency diffusion, but they end at the ceiling line. The ceiling was really a curtain of glowing LEDs that resemble a star field, and above that are flat masonry walls triple painted to seal the block. The ceiling structure, drape pockets, forestage grid, and large round return air ducts provide diffusion in this zone. Shifting from wall treatments below the ceiling to suspended-in-space diffusion elements above was a potent and cost-effective strategy.

Ceiling reflectors are another area where only the minimal ceiling was used—no surface was there simply for architectural form or conceit. In fact, the ceiling is a netting of LED lights that is acoustically transparent, hung from the catwalks. The only ceiling that is needed acoustically is the forestage array that ends at the edge of the stage extension and the ceiling below the catwalk over the balcony.

Mark Holden
November 9, 2011

CASE STUDIES

V

Appendix I: Alice Tully Hall, Lincoln Center for the Performing Arts, New York, NY: From Dry Hall to Warm, Glowing Vessel

Introduction

The building that houses Alice Tully Hall (ATH) and The Juilliard School was designed by architect Pietro Belluschi and opened in 1969. It is named for Alice Tully, a New York performer and philanthropist whose donations assisted in the construction of the hall. This hall was the first major concert venue in New York City designed specifically for chamber music, and the last building was completed within the original Lincoln Center for the Performing Arts complex. A concert in 1969 marked the first performance by the new chamber music organization that would make the hall its home, the Chamber Music Society (CMS) of Lincoln Center. In October of 1969, The Juilliard School celebrated its opening with three concerts in ATH (see Figure A1.1).

Since the hall's opening, ATH has been utilized for programming including the Mostly Mozart Festival, the Great Performers series, and the Lincoln Center Festival and has served as the venue for a jazz series. The hall has been the setting for thousands of special events including performance debuts, world premieres, star-studded openings, and galas such as the annual New York Film Festival. The hall was used year-round and held approximately 750 annual events. After nearly 40 years, the hall was temporarily closed for renovations following a "Good Night Alice" gala and concert on April 30, 2007.

Building Blocks

In 1998, Jaffe Holden was hired to survey all of the performance spaces at Lincoln Center as part of a master plan study spearheaded by Beyer Blinder Belle Architects & Planners. The scope of

Figure A1.1 Alice Tully Hall, New York, NY, renovated 2009. Wood veneer from a single tree wraps the interior of the hall. Sidewall "fins" direct early reflections into the central seating area. Extra wide seating rows provide comfort seldom found in New York halls. Seats are custom-designed for minimal sound absorption and are covered in fabric found in exotic automobiles. (Photographed at Lincoln Center for the Performing Arts in New York City.)

this study was to determine how best to upgrade the campus, which had not had a comprehensive renovation in nearly 30 years. A complete set of campus drawings did not exist, since each of the four main structures was built by different architects at different times. The four main structures are the Avery Fisher Hall, formerly called Philharmonic Hall, the Metropolitan Opera House, the New York State Theater (now named the Koch Theater), and The Juilliard School.

In 2007, Diller Scofidio + Renfro (DS+R) in collaboration with FXFowle Architects began a billion-dollar renovation, which included public spaces, plazas, circulation, and support areas, as well as Juilliard and ATH. See *Lincoln Center Inside Out* (Diller, Scofidio, and Renfro 2013), for a visual description of the entire project.

Challenge

Lincoln Center wanted to protect the integrity of the hall's acoustics. I felt that the hall needed a significant acoustic overhaul.

Solutions

By optimizing the hall's side wall–shaped stage ceiling and adding adjustable acoustics and a new stage extension, the acoustics were vastly improved.

Design Process

Acoustic Survey

Before the renovation could begin, I needed to complete an acoustic survey—the first since its opening. I was no stranger to this hall. I had attended concerts by the CMS of Lincoln Center and felt that improvements could be made to the sound of chamber music. I remember hearing Mozart in a noisy hall with NC levels of 25 plus low RT and G levels. It was like listening to a good stereo recording of music from speakers on the stage without enough room support. There were ample leg room and comfortable seats, but the hall was dimly lit and needed more visceral energy and connectedness between the performers and the audience.

During my first survey visit, there was no chamber music performance, and the hall looked completely different from what I recalled. The stage was decked out in dozens of moving lights, video projections were on the upstage walls, and a fashion runway stage was set up over the front rows of seats. I was told that there was a rehearsal for a product rollout by a major cosmetics firm. Bass from substantial stage-mounted subwoofers filled the hall with a controlled pounding rhythm that surprised me; I had expected a poor bass sound that was boomy and uncontrolled. Was the sound clean because the low-frequency RT and BR were low? Most recital halls have a BR of 1.1–1.2 and require substantial adjustable acoustic drape or banner systems to absorb excess energy. I knew that this hall had no such system, and I found the clean bass sound to be odd in nature.

The stage was also configured differently from what I recalled. Wood stage side walls, actually huge doors, extended from floor to ceiling. These were swung open to create side stage wings. I had recalled the ceiling over the stage to be solid wood, but now it was filled with theatrical lighting instruments pouring light through large openings. This was like no recital hall I had seen before.

Acoustic Mysteries Uncovered

Rumble

I discovered a few acoustic mysteries during my exploration of ATH. Other than carpet and seat replacements, the hall appeared much as it had in 1969. It had rich, red seats and vertical mahogany wood battens on every wall and ceiling surface. A flexible, reconfigurable stage was a remarkable design feature especially in the 1960s. The facility was the most used facility on campus for events including Juilliard and CMS concerts, film premieres, red carpet film awards, network television pilot showings, jazz concerts, and church service every Sunday morning using the hall's pipe organ.

The first discovery was that the hall had a very noisy HVAC system. Our review of the original design documents for the air systems revealed that high-velocity air in unlined supply ducts projected air to patrons through slot diffusers along the ceiling. Air was returned through similar slots along the ceiling plane between the wood battens. Acoustic measurements confirmed that the air velocity noise produced noise levels of NC-24–30 from units next door to Juilliard's main mechanical room. The units ran 24/7 for nearly 30 years. In fact, it was not clear how to shut off the units.

After de-energizing the air units, there was a fantastic sense of quiet and serenity in the hall as the noise dropped from NC-17 to NC-15. Remediating velocity noise is straightforward, and I believed that our work to improve this aspect of ATH could be easily accomplished. Hopes for a quick solution were dashed when we heard the subway rumble from the Number 1 and 2 express trains on the Broadway line. My heart sank. Subway rumble had never been mentioned by staff or users during our interviews. It had been masked by the loud ambient air noise for years. When the air velocity noise issue was solved, the audibility of the subway rumble was increased.

Our measurements provided clarity. The subway rumble with maxima in the 125-Hz octave band was masked by air noise at 250 Hz and above. An interesting condition was occurring where masking at higher frequencies actually covered low-frequency noise, even though the air noise levels at 125 Hz were less than the subway rumble. I recall that I was greeted with disbelief when I told the design team and the Lincoln Center redevelopment team (our client) that we had a serious subway rumble issue. I brought the team to ATH and had the chief building engineer shut off the air and demonstrate the rumble. More than one constituent remarked, "Now that you have pointed out something I haven't heard for years, I will never be able to listen in this hall again without hearing that sound." I had allies in the quest to make this hall one of the quietest halls in New York.

Wall Absorption

The last mystery required significantly more sleuthing on my part. The acoustic modeling predicted a mid-frequency RT of about 1.5 seconds based on the volume and surface materials. We expected that the calculation might overestimate the RT slightly as we cannot account for the drying of the mahogany wood walls, cracks in the paneling, duct openings, and so on. Yet, I was shocked that the actual measurements of RT in Tully were much lower at about 1.2 seconds especially at the low end. After repeated recalculations to prove that no mistakes were made, I suspected that the wood walls were causing Tully to be so dry.

After careful study of the drawings and much observation, we determined that the walls were made of 2 × 2-in. (51 × 51-mm) mahogany wood battens spaced 2 in. on the center. These were screwed to a black cloth covering 3/4-in. (19-mm) plywood backing mounted on wood furring strips on top of block walls. This construction was confirmed through openings found in the walls at electrical boxes. This could not be correct, or else Sabine's formula had been violated, and the laws of physics were rescinded. Stage hands and architects were certain that this construction was consistent on all of the walls.

I requested a ladder from the stage crew and climbed 30 ft. to the upper side walls. I had a hunch—the walls looked identical from below, but were they?

With my flashlight (required equipment on all site visits) and pen, I probed the spaces between the battens and found them to be soft. Black fabric concealed 2-in. fiberglass batts in large zones of the upper wall surfaces. In addition, I found plywood on the upper parts of the wall; it was edge supported and only 1/4 in. (6 mm) thick. It was a tuned bass panel absorber. When we factored these surfaces into the model, our calculations were right on target.

Stage Extension Mock-Up

In January 2005, acceptance was needed from constituents for the idea of a stage extension. There was concern about extending the stage and positioning the musicians far out in the hall. We were confident that reducing the seat count and removing seating absorption from the room would improve RT. The stage extension would move the source position closer to the hall's waist where the walls flared back would improve D50 and C80 reflections.

Extending the stage was also critical to improving intimacy and contact, but this controversial move would need to be proven. I proposed to demonstrate the acoustic feasibility of the stage extension with a physical mock-up and with full participation of the CMS.

A rented portable stage was built over the top of seats to simulate new wood stage extension, and the Emerson String Quartet was commissioned to play in multiple positions for evaluation. In order to document the results of this one-time-only test, a questionnaire was prepared for musicians to register opinions about the quality of sound and the ease of onstage hearing. An invited audience filled out similar questionnaires describing the quality of chamber music when the quartet moved forward on the stage extension. The results were clear. Invited guests were unanimous in their opinion that the sound with stage extension was louder, had more impact, and had a longer RT. Musicians were divided on if it was easier to hear each other on the extension. One musician quipped, "Anyway, we never listen to each other!"

Renovation Begins

With the survey complete, the renovation to the interior could begin.

Interiors

Due to the accelerated 20-month schedule and budget limitations, the structural bones of the hall such as the balcony, floor, and ceiling needed to remain in place, whereas the interior skin was reimagined by the architects and theatre consultants FDA to be a glowing vessel that surrounded the performers and the audience. Central to the design was the skin (the walls), which would glow in a warm blush when backlit.

The interiors need to be solid and substantially supported to create the acoustic reflections and strong reverberation preferred for groups such as the CMS of Lincoln Center and the Juilliard Orchestra. Materials that glow tend to be thin and unsubstantial, such as glass or plastic, and are mid and low frequency absorbent. With the New York subway tracks only yards away, the new interior skin would need to be a box-in-box construction such that 63–500-Hz vibration levels from the metal wheels on the subway tracks would not be telegraphed into the renovated hall.

Our challenge was to determine how this new skin could be both glowing from behind and be solid and substantial. We needed to determine how the recommended acoustic reshaping could be incorporated into the theme and be within budget. In addition, our challenge was to mount the walls such that the vibration levels would not exceed NC-17 yet be solid enough to reflect sound efficiently.

The solution required a truly collaborative design process with the architects, lighting designers, acousticians, structural engineers, and manufacturers. Scheme after scheme was put forth and mocked up with different materials, and each scheme was reviewed and refined by team members to achieve an acoustically workable and cost-effective solution that met all needs. Schemes needed to be reviewed by all the stakeholders and user groups, as well as the multiple boards and The Juilliard School.

To achieve the warmth and richness of wood and to get the wood to glow, a thin veneer of wood was imbedded in a 1 in. (25 mm) thick, heavy clear resin by 3form and lit with colored LEDs from the rear. Other wood wall panels were formed of heavy 1 and 1/2 in. (25 and 13 mm) thick medium-density fiberboard (MDF) with the wood veneer applied.

Wall Shaping

The new walls of the hall have been reshaped into sinuous curves to optimally distribute sound reflections and sustain a bright, clear sound for classical concerts, recitals, and chamber music. The walls now splay in a way to achieve more optimal C80 reflections and create diffusion surfaces by the aggressive articulation (Figures A1.2 through A1.4).

The 3form resin was unlike any material before used in the design of extensive surfaces in a concert hall. This led to serious concerns regarding the acoustic performance of the material on the part of the constituents and designers, as well as our team. The material density was similar to heavy plaster, but we wondered if the crystalline resin structure would be absorbtive at definable frequencies and if the shadowless mounting details and structural supports for backlighting would increase mid- and low-frequency absorption. A mock-up panel was built to match the new proposed design and tested in an acoustic lab. The results were encouraging.

Stage Extensions

Based on the mock-up, we moved ahead with designing a stage extension with FDA. To create a more intimate musical environment, two automated stage extensions would allow for adjustable staging options and audience capacities. New stage ceilings over the musicians would be tuned to enhance onstage hearing and projection of sound to the audience. Acoustic banners would be dropped from the ceilings for amplified sound for jazz concerts or new music festivals, and speaker systems behind the portable movie screen and surrounding the walls would support top-quality film sound.

Solutions for Subway Noise and Vibration

Due to the hall's proximity to the subway, improving isolation for recording, rehearsal, and performances had been a top priority. For years, the subway noise in the hall was almost inaudible

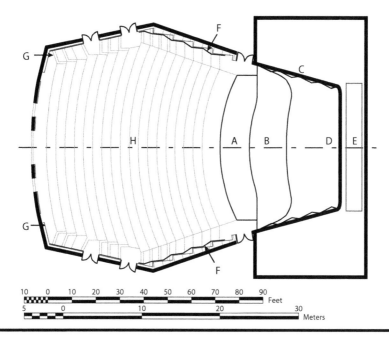

Figure A1.2 . Alice Tully Hall, New York, NY, renovated 2009. Orchestra plan. (A) Stage extension. (B) Orchestra pit lift/stage extension. (C) Rotating stage walls float on rubber isolators. (D) Rubber isolators support rear wall that slides into storage to expose pipe organ, (E). (F) Reshaped sidewalls for improved lateral reflections include LED backlighting. (G) Rear walls and sidewalls float on rubber isolators. (H) New concrete floating floor supports seating area.

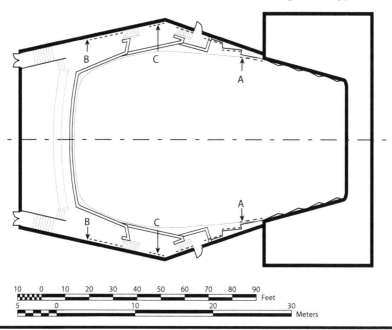

Figure A1.3 Alice Tully Hall, New York, NY, renovated 2009. Balcony plan. (A) Double-layer wool serge acoustic banners in front of reshaped walls. (B and C) Acoustic banners over resiliently supported rear sidewalls. Balcony floor was not floated due to structural restrictions.

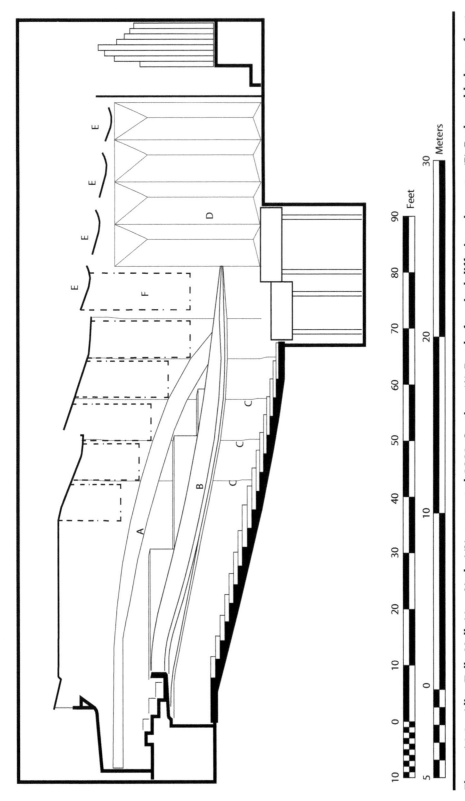

Figure A1.4 Alice Tully Hall, New York, NY, renovated 2009. Section. (A) Band of conical diffusion elements. (B) Reshaped balcony front. (C) Lower sidewalls shaped for lateral reflections and diffusion. (D) Stage walls can rotate 180°. Bottom portion can rotate to make entry door. (E) Airfoil shaped ceiling reflectors can be tuned. (F) 12 sidewall acoustic banners.

due to noisy air systems (NC-25–30) in the hall that masked the rumble (NC-27). After the engineers shut off the noisy air systems, we demonstrated to the team and the stakeholders that with a quiet air system, the subway would actually be more noticeable. As a result, the solution would require a complex box-in-box isolation scheme for the walls and the floor. To support this effort, we benchmarked noise measurements of other halls in New York and how they were in fact much quieter than ATH.

After costing such isolation systems, it was determined that our recommended box-in-box construction of the hall and on stage was cost prohibitive. Undaunted, we suggested a lower-cost option to partially box in box the room, treating only those surfaces that audibly radiated subway noise. We then asked the Metropolitan Transit Authority (MTA; who owns and operates the subway) to get the noisiest rails mounted on modern rubber pads as well as weld the joints between the rails to limit the source of the vibration. We were told that this would never happen.

The team got a lucky break when it was discovered that a new Lincoln Center board member was also on the MTA board. The board member was able to convince the MTA to advance the schedule for upgrading the old and noisy track.

Architectural Details

Interior Wood

Opinions differed about the acoustical quality of the original hall. The mandate from Lincoln Center was to do no harm acoustically and make modest improvements where possible. DS+R and FXFowle's vision was to grab the patron's attention with the undulating interior of Moabi wood as it wrapped the hall in a remarkably complex pattern. The effect of the wood would be so warm and seamless that it would be like sitting inside a musical instrument. Astonishingly, the wood would come entirely from one tree. The Moabi was harvested from a single, 40-ft. tree in Gabon and transferred to Japan where it would be sliced and heat-bonded to resin or MDF panels.

Adjustable Acoustic Systems

In order to make the space suitable for various types of events, especially for film, a set of acouStac banners from Pook Diemont & Ohl was installed. The banners weigh 1300 lbs. (590 kg) and are able to travel 150 ft. (46 m)/minute. There are 18 acouStac banners of variable lengths lining the hall's side wall and drop from slots in the ceiling (see Figures A1.5 and A1.6). Rotating the side walls onstage and deploying these banners result in a very good room for amplified sound and film.

We worked with the architects to optimize the surface acoustically, making the sound significantly clearer and more present. This was achieved through shaping and restructuring the interior skin. With the wood's many twists and folds, the walls now reflect sound with the appropriate frequency response and in the right direction. A distinctive feature of the upstage wall is an array of dots. The dots are large in the center and gradually grow smaller as they fan out in a pattern that is suggestive of pop art. These are conical sound diffusors that mellow some of the harshness and brightness that might occur from a flat upstage wall (Figure A1.7). A similar pattern of dots formed of GFRG wrap the hall at the transition from the dark drywall walls to the wood surfaces. While the amount of diffusion in the hall was a subject of debate and concern, especially from peer

Figure A1.5 **Alice Tully Hall, New York, NY, renovated 2009. Partially deployed acoustic side-wall banners are used to control unoccupied RT in rehearsal mode. The Juilliard Orchestra uses this setting and rotates the upstage walls to vent energy on stage. (Photographed at Lincoln Center for the Performing Arts in New York City.)**

reviews, the combination of the conical diffusers and the bends in the wall and ceiling surfaces is perfectly adequate (Figure A1.8).

The glowing side walls of the stage also rotate and spin, and the back side of each wall has a black sound-absorptive material. Mounted over large air cavity, the back side of the walls is an excellent bass absorber. In effect, these are tormentors used for masking for film screenings and other amplified productions (Figure A1.9). The upstage wall also contains six doors that open to reveal Tully's pipe organ.

It was vital to effectively isolate the room from the outside world as well as The Juilliard School above and around the hall (Figure A1.10). To deal with subway noise, as stated in the section Solutions for Subway Noise and Vibration found in this chapter, the MTA was persuaded to weld the nearby train tracks. Rubber pads were also added under the rails. The HVAC system was improved, and velocities were significantly lowered. The air supply and return all came from above because drilling holes in the existing slab for an underfloor supply system would have caused the old floor to become structurally unstable and collapse. A rating of NC-17 was achieved despite the main mechanical rooms being immediately adjacent house left.

Custom-built acoustic isolation doors were added to the auditorium. The doors are sealed to the walls in a unique frameless way where no hardware was visible, and magnetic seals hold them closed like a refrigerator door. We replaced all the surfaces leading down to side galleries in the inner lobby with 1.5 in. (38 mm) thick gray felt in an interesting overlapping pattern. There is also a gray

Figure A1.6 Alice Tully Hall, New York, NY, renovated 2009. For amplified events, acoustic banners are fully deployed. DS+R specified high-speed motors so banners move very quickly. They are often moved during shows. (Photographed at Lincoln Center for the Performing Arts in New York City.)

Figure A1.7 Alice Tully Hall, New York, NY, renovated 2009. Wood conical diffusers on upstage wall vary in depth and diameter providing high-frequency diffusion. Upstage walls can slide into pockets revealing pipe organ. (Photographed at Lincoln Center for the Performing Arts in New York City.)

Figure A1.8 Alice Tully Hall, New York, NY, renovated 2009. Stage walls shaped for sound projection and diffusion are back-lit with LEDs. Both stage lifts are in raised position for Chamber Music Society of Lincoln Center performance. Airfoil shaped ceiling reflectors include orchestral lighting but can tip 90° to reveal motorized rigging and more extensive theatrical lighting. (Photographed at Lincoln Center for the Performing Arts in New York City.)

Figure A1.9 Alice Tully Hall, New York, NY, renovated 2009. Stage wood walls rotate to expose black fabric-covered acoustic panels with large air cavity for low-frequency absorption. Horizontal slots ventilate LED glow lights. (Photographed at Lincoln Center for the Performing Arts in New York City.)

matching carpet. It creates a sound lock zone that is very sound absorptive. We advocated to the architect that it would be ideal for audiences to undergo a kind of acoustical order of events as one enters ATH from the outside. First, the patron moves from the outside lobby, which is bright and reverberant, to the inner lobby, which is quieter with acoustic plaster ceilings. Then, the patron advances into the side galleries, which are very quiet, and then into the sonically live hall. It is a procession designed to give a sense of surprise to the ear.

Measuring Results

The stage extension survey indicated a strong preference for the downstage position and a moderate preference for screens behind the musicians. This improvement was consistent across all eight of the quality criteria. Guest evaluators were positive in their response to a forward placement of the ensemble. They seemed unanimous in opinion that the sound had more impact, reverberation, and intimacy.

In 2009, we were able to enter the hall and take a full series of acoustic measurements. We were able to measure the hall in all modes, from fully reflective (all banners stored) to all banners deployed and stage panels rotated. The graphs in Figures A1.11 and A1.12 show unoccupied results.

Acoustic Tuning

When the hall was completed in January of 2009, we requested that the CMS assist in tuning the hall's ceilings over the stage. They aided in determining the sweet spot on stage for solo and chamber performances. A schedule was finalized, and tuning went forward with great success.

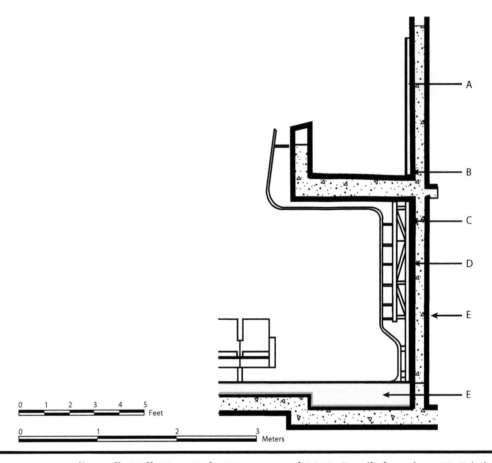

Figure A1.10 Alice Tully Hall, New York, NY, renovated 2009. Detailed section. (A) Existing concrete floor structure. (B) Vibration isolators. (C) Wood floor on floating concrete slab. (D) Wood veneered surface on either resin or MDF. (F) Steel frame supports wood skin and LED backlighting, lighting on resilient mounts. (G) Suspended wood soffit. (H) Three-layer drywall wall on resilient mounts.

The results of the tuning were so favorable that Lincoln Center decided that it was worth the risk to invite *The New York Times* music critics to the second round of acoustic tuning. The results appeared on the front of the *Times'* arts section the next morning: "Musicians Hear Heaven in Tully Hall's New Sound" (Wakin 2009).

Press

"Oh my God, it's heaven," said Anne-Marie McDermott, the pianist, after playing a Steinway on the stage. "You can do anything: the clarity, the range." She called the sound fat, rich and buttery, and unfamiliar from pre-renovation days. "I wouldn't have recognized it," she said.

With giddiness and glee, musicians tested the acoustics of the newly renovated Alice Tully Hall on Tuesday, less than a month before it reopens after a $159 million, 22-month upgrade, a major milestone in Lincoln Center's $1.2 billion remaking. "I'm already so

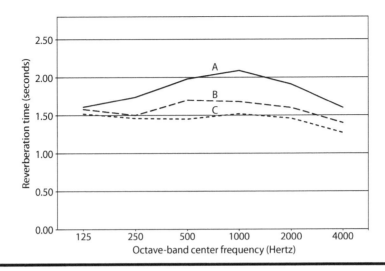

Figure A1.11   Alice Tully Hall, New York, NY, renovated 2009. Reverberation time. (A) Banners fully stored. (B) Banners partially deployed. (C) Banners fully deployed and stage walls rotated.

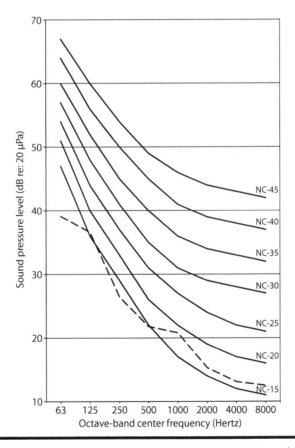

Figure A1.12   Alice Tully Hall, New York, NY, renovated 2009. Target of NC-17 was achieved with box-in-box construction for floors and walls to isolate subway. Air systems defy convention with both supply and return air from the ceiling, yet are noiseless.

jazzed," said David Finckel, a cellist and co-artistic director of the Chamber Music Society of Lincoln Center, Tully's main inhabitant. "It's the greatest new toy I've ever had."

Daniel J. Wakin
The New York Times (2009)

Project Information

DESIGN TEAM

Architect: Diller Scofidio + Renfro
Theater Consultants: Fisher Dachs Associates
Architect of Record: FXFowle Architects

DETAILS

242,000 SF
1100-seat theater

BUDGET

$180 million

SCOPE OF WORK

Architectural acoustics
Sound isolation
Mechanical system noise control
Full-featured sound reinforcement system
Adjustable acoustics

CLIENT

Lincoln Center for the Performing Arts

COMPLETION DATE

2009

Schedule for Alice Tully Hall Acoustic Tuning

The following tuning checklist was prepared by myself and Larry King, Project Manager, for use in tuning the hall with musicians and representatives from Lincoln Center for the Performing Arts (LCPA). It is listed here as an example of the detailed planning required for a successful tuning process.

January 20, 2009

1:00 p.m.–2:00 p.m. Lincoln Center for the Performing Arts (LCPA)/Jaffe Holden (JH) Setup

Tour hall and backstage spaces; check them for proper operation of doors, lights, toilets, heating, ventilating, and air-conditioning systems; set up the talkback sound system and intercom for ease of stage-to-house communications. Piano will have been tuned? LCPA setup—security, waivers, schedule, refreshments, etc.?

2:00 p.m.–3:30 p.m. Stage Setup

■ Set fixed stage configuration: both chair wagons on lifts
■ Test winch controls for moving acoustic ceiling panels and acoustic banners
■ Set performer stage accommodations—cello riser(s), chairs, music stands, piano, and bench (page turner?)

3:30 p.m.–4:00 p.m. Perform Tests with Fixed Stage

■ Three pauses of 1 minute each to adjust acoustic ceiling panels
■ Works to be performed, in sequence (LCPA)

4:00 p.m.–4:30 p.m. Changeover

■ Fixed stage to extended stage
■ Fill out listener and performer questionnaires
■ Discussions with constituent spokespersons and performers

4:30 p.m.–5:00 p.m. Perform Tests with Extended Stage

■ Three pauses of 15 minutes each to adjust acoustic ceiling panels
■ Works to be performed, in sequence (LCPA)

5:00 p.m.–5:30 p.m. Changeover

■ Extended stage to thrust stage
■ Fill out listener and performer questionnaires
■ Discussions with constituent spokespersons and performers

5:30 p.m.–6:00 p.m. Perform Tests with Thrust Stage

■ Three pauses of 15 minutes each to adjust acoustic ceiling panels
■ Works to be performed, in sequence (LCPA)

6:00 p.m.–6:30 p.m. Break

■ Quick meals?

6:30 p.m.–7:30 p.m. Continue Tests with Thrust Stage

■ David Finckel and Wu Han
■ Escher Quartet arrives at 7 p.m.
■ Three pauses of 15 minutes each to adjust acoustic ceiling panels
■ Works to be performed, in sequence (LCPA)

7:30 p.m.–8:00 p.m. Changeover

- Thrust stage to extended stage
- Fill out listener and performer questionnaires
- Discussions with constituent spokespersons and performers

8:00 p.m.–9:00 p.m. Perform Tests with Extended Stage

- David Finckel, Wu Han, and Escher Quartet
- 3 pauses of 15 minutes each to adjust acoustic ceiling panels
- Works to be performed, in sequence

9:00 p.m.–9:30 p.m. Changeover

- Extended stage to fixed stage
- Fill out listener and performer questionnaires
- Discussions with constituent spokespersons and performers

9:30 p.m.–10:00 p.m. Perform Tests with Fixed Stage

- David Finckel, Wu Han, and Escher Quartet
- Three pauses of 15 minutes each to adjust acoustic ceiling panels
- Works to be performed, in sequence

10:00 p.m.–11:00 p.m. Wind Down

- Wrap up discussions
- Collect questionnaires

Appendix II: Baldwin Auditorium, Duke University, Durham, NC: A Domed Homecoming

Building Blocks

My first encounter with Baldwin Auditorium was in 1977 as a Duke undergraduate student when I was working for Student Services, the in-house AV group at the university. I was sent there with a couple of Schoeps microphones and a Nagra reel-to-reel tape recorder to record a string quartet performance. I cannot vouch for the recording quality, but I do recall that the hall was worn out and in poor condition. The acoustics of the room were boomy, with a very high BR and low- to mid-frequency reverberation time (RT) probably due to the vintage acoustic tile glued to the walls and ceiling. The air-conditioning was so noisy that it had to be shut off during the recording session.

I returned to Baldwin in 2010 to conduct an acoustic survey of the hall as part of a new commission to evaluate the Nelson Music Room and the auditorium for a possible renovation of the music department. The room looked and sounded very much as I remembered, but a new stage extension had been added to expand the performance platform. I observed the Duke Symphony Orchestra tightly crammed onto the stage, surrounded by rolling shell towers only 12 ft. (3.5 m) high. Overhead in the small stage house were three acoustic ceiling shell pieces that were oddly angled, sending all sound out to the hall (see Figure A2.1).

Measurements confirmed the reported dryness of the hall, which was less than 1.5 seconds RT mid frequency and even less at the high end. The walls under the acoustic tile were made of thick plaster on masonry, and the ceiling was made of plaster on suspended metal lath. In the center of the hall was the iconic coffered dome. The dome was huge, encompassing almost all of the ceiling area, and was formed of thin plaster in a structure similar to an egg shell. The dome was extensively coffered with ribs that provided mid- and high-frequency diffusion; however, acoustic tile had been affixed to the center of each coffer with asbestos glue. The hall's poor acoustics were often blamed on the dome. I heard a number of comments about how the sound was "lost" in the dome and how it focused the sound in undesirable ways. However, I was not so sure that the dome was the real issue here. Perhaps it was an opportunity.

Figure A2.1   Baldwin Auditorium, Duke University, Durham, NC, 1930. East Campus's Baldwin Auditorium, pre-renovation. This photo was taken in 2009 and the hall is unchanged since a 1960s renovation. Note the wood seats, single pane windows to the exterior, peeling paint, and large domed ceiling.

I was more concerned with the extensive width of the seating area. The hall was a good deal wider than it was deep, which is not good for acoustics. We wanted to provide early lateral reflections for the seating areas, but the 120 ft. (36.5 m) wide room would be an issue for any renovation. Many rows under the balcony suffered from the low ceiling, and worse still, outdated and noisy blowers were located in the plenum below the seats. These units were so noisy that they needed to be shut off during symphony performances or recording sessions, which prevented ventilation. The final flaw was the lovely arched windows that flooded the room with natural light—they let in a lot of noise from the outside.

The university's goal was to transform the outdated auditorium into a modern multi-use concert hall that would serve as a home to the renowned artists of Duke Presents and the school's music department. The Emerson String Quartet was a regular visitor as were leading solo artists like pianist Emanuel Ax. Amplified events such as jazz concerts and large lectures using a projection screen were also part of the programming mix. Any event but fully staged opera or musical theater was possible, although music events would dominate the program.

Challenge

The challenge was how to preserve the wonderfully historic dome and yet significantly improve the acoustics for the full range of uses. Also, the stage was partially located in the stage house and the hall. How could they be acoustically coupled with a unifying acoustic shell?

Solutions

The dome was treated with diffusive panels to transform it from a liability to an asset, while a dynamic new wood shell surrounded the stage and extended dramatically into the hall. Both rolling acoustic banners and horizontally tracking acoustic drapes were utilized (Figure A2.2).

Creating the Building

The design team featured Bill Murray from Pfeiffer Partners, Robert Long of Theatre Consultants Collaborative, and myself. Our initial acoustic design was a significantly smaller beautiful new 685-seat hall within the Baldwin interior. Downsizing the seating was critical to the success of the design (Figures A2.3 and A2.4). The volume of the hall was low for symphonic reverberation at over 1000 seats but could be workable with a smaller number. This reduction allowed the width of the hall to have much improved C80 reflections. A shallow wraparound balcony with soffits would direct the reflections downward into the orchestra seating area (see Chapter 11 for information on balcony reflections), add intimacy and interest to the very wide room, and provide diffusion opportunities (Figure A2.5).

Architectural Details

Ceiling

The existing dome was to remain, but the university and its project manager were concerned about the negative acoustic effects. We believed that the dome, minus the glued-on tile, would be a good diffusive device and would add critical acoustic volume and diffusion in the otherwise plain room. Standing directly under the dome, there were certainly some focusing effects, but we advocated that the dome was a good thing that should remain. If direct sound from the stage could be blocked from impacting the dome with a large forestage ceiling, it would add reverberation and diffusion. The remaining flat plaster areas of the ceiling that were covered in mineral acoustic tile would be stripped back to the original plaster for full-frequency reflections. The challenge remained of how to utilize the beneficial volume of the dome and its diffusion properties while eliminating the focusing effects.

Careful analysis of the dome's geometry and ribs convinced us that the best direction would be to strip the tile back to the plaster and add thin RPG BAD diffusive panels to the rear-focusing area coffers. To block direct sound from the stage from ever reaching the dome, we proposed an extensive wood acoustic forestage reflector covering the new enlarged stage extension. This reflector faithfully followed our acoustic guidelines for sound projection, sound diffusion, and onstage hearing. It included openings and gaps between sections that would allow sound to reach the upper reverberant volume of the hall. A newly designed, gorgeous wood ceiling and walls extended into the stage house volume.

The wood stage enclosure features double doors that allow for the free movement of performers, openings for stage air distribution, stage lighting, and line array speaker suspension points (Figure A2.6). The complex diffusive geometry was built of 1 in. (2.5 cm) thick double-curved wood by a local furniture maker. It was built in large sections then delivered and hoisted into position. The tight budget would not allow the inclusion of motorized winches to allow adjustability

Figure A2.2 Baldwin Auditorium, Duke University, Durham, NC, renovated 2013. This renovation by Pfeiffer Partners architects preserves the character of Baldwin Auditorium but enhances the intimacy, comfort, and acoustics.

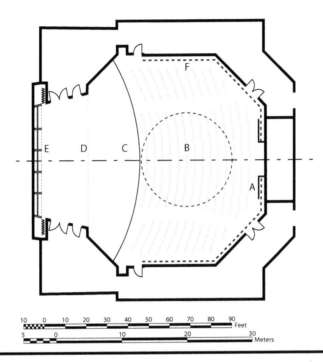

Figure A2.3   Baldwin Auditorium, Duke University, Durham, NC, renovated 2013. (A) Motorized acoustic drapes pockets near control booth. (B) Location of domed ceiling overhead. (C) Stage extension with perforated floor. (D) Line of proscenium opening. (E) Shaped wood rear wall with acoustic drape. (F) Acoustic drapes on sidewalls have cut-out for exit doors.

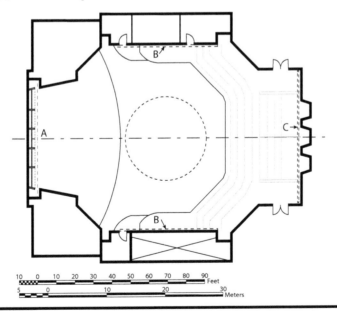

Figure A2.4   Baldwin Auditorium, Duke University, Durham, NC, renovated 2013. (A) Manual pull acoustic drape on upstage wall stores in side pockets. (B) Roll-down acoustic banners on sidewalls cover diffusive wall with small glazed areas. (C) Rear wall acoustic banners pull up from banner boxes.

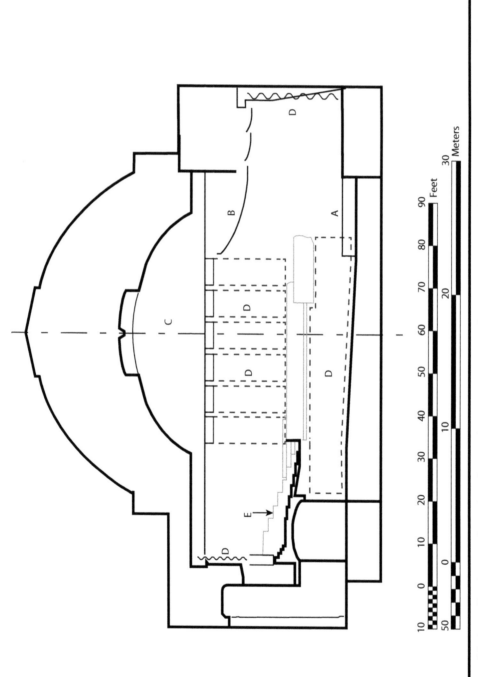

Figure A2.5   Baldwin Auditorium, Duke University, Durham, NC, Renovated 2013. (A) New wood stage extension with air perforations (see Chapter 10). (B) Wood forestage reflector (see detail 13.6). (C) Domed ceiling with applied diffusion panels. (D) Adjustable acoustic drapes and banners. (E) Balcony build-up similar to detail 14.5.

Figure A2.6 Baldwin Auditorium, Duke University, Durham, NC, renovated 2013. Wood veneer plywood and MDF walls and ceilings are shaped for acoustic reflections, diffusion, and blending. Openings in the ceiling reflectors are precisely positions for acoustic coupling to the volume above them.

of the angle or height of the ceiling elements. We would have one shot at getting the angle right, and it would need to be in the design phase.

Adjustable Acoustics

The location of the acoustic banners and drapes presented a challenge, as we did not want to penetrate the ceiling with banner boxes on account of access issues and limited space for the mechanisms above the ceiling. Therefore, the banners were placed on the side and rear walls in visually exposed banner boxes. The new side walls held more than just the banners. They also narrowed the hall to 90 ft. (27.5 m) for improved C80 reflections and diffused sound with applied painted wood panels, vision lites, and moldings. The banners are double-layer wool serge (as described in Chapter 15) and drop down on double rollers to minimize the size of the exposed storage enclosure. The rear wall velour drapes pull up from floor-mounted banner boxes (Figure A2.7).

A motorized traveler drape can cover the walls under the balcony on either side of the control booth. This is useful for controlling reflected sound from amplified events. A large manual acoustic drape covers the upstage wood wall for amplified productions and lecture use and can be stored inside pockets when not needed.

Figure A2.7   Baldwin Auditorium, Duke University, Durham, NC, renovated 2013. A view of the acoustic banners deployed on the upper sidewalls and tracked drapes under the balcony.

Stage Floor Air

The HVAC engineers were concerned that the under floor air supply in the auditorium seating area needed to continue into the stage in order to provide appropriate comfort on stage. Typical round metal floor supply registers that could be readily hidden under the audience seats were rejected for use in the stage floor as they were unsightly and got in the way of performances. Our innovative solution was to drill 75,000 small air holes in the stage floor—small enough to be virtually invisible but large enough for the air to filter up and cool the musicians above (Figure A2.8). An air plenum was built below the stage floor and was pressurized by slow-moving, muffled conditioned air. Mock-ups proved the system's feasibility mechanically and acoustically. The system is rated below NC-15, and the final result is detailed in the drawing in Figure A2.9.

Measuring Results

The design team's approach for the renovation of Baldwin Auditorium was an acoustic triumph for all types of programs, ranging from amplified jazz and African drumming to solo piano and voice recitals. Praise from artists, audiences, and presenters was unanimous. Besides being visually stunning, the strength of the RT, especially at low frequencies, is excellent. We were concerned that the thin plaster of the dome would lower the BR through diaphragmatic absorption. However, this did not turn out to be the case probably because of the dome's rigidity and balance of massive materials on the remaining surfaces.

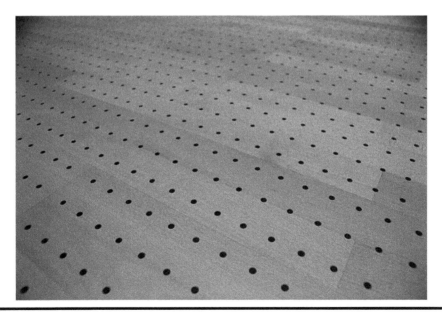

Figure A2.8 Baldwin Auditorium, Duke University, Durham, NC, renovated 2013. Air conditioning on stage extension is provided by an innovative perforated wood floor. Cello players remarked that the holes are ideally sized for their pegs.

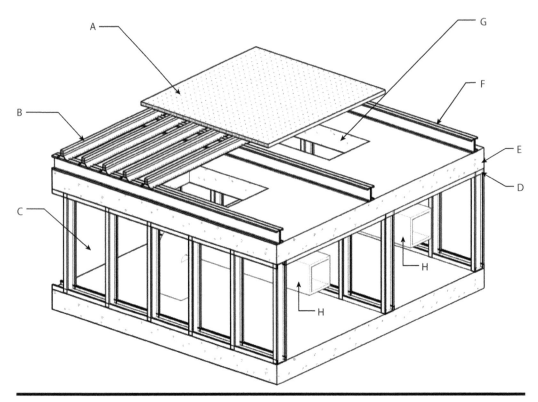

Figure A2.9 Baldwin Auditorium, Duke University, Durham, NC, 2013. Stage floor detail. (A) Perforated T&G wood floor on subfloor. (B) Support channels. (C) Acoustic air supply plenum. (D) Acoustic duct liner. (E) Structural floor slab on metal decking. (F) Back-to-back metal studs. (G) Air openings. (H) Supply air ducts.

Settings for Adjustable Acoustic Systems

We provided an acoustic manual for this project to outline recommended settings of the adjustable acoustic elements for various types of programs in the auditorium. Note that all of the settings for the acoustic drapes and banners listed here have been recorded as presets in the rigging control module and are labeled accordingly for ease of use. In addition, performer stage locations were experimented with throughout the tuning process, and the approximate recommended locations are described. For programs not included, the variable acoustic elements would be set and the performers located according to the event type that they most closely resemble (Figure A2.10).

Explanation of Acoustic Settings

The following excerpts are from the actual user's manual for Baldwin Auditorium. The images of musicians' recommended positions on stage have been omitted.

1. Symphony/chamber
   a. This setting would be used for most large acoustic ensemble performances including the university's symphony orchestra, wind symphony, choral, and chamber ensembles.

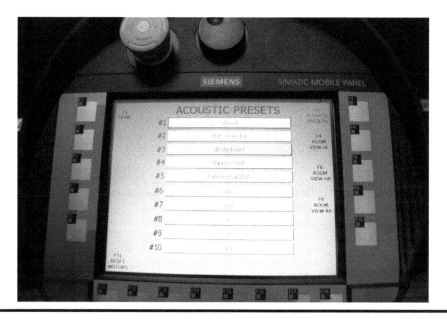

Figure A2.10   Baldwin Auditorium, Duke University, Durham, NC, renovated 2013. Adjustable acoustic control system has presets predetermined during our acoustic tuning. This allows users to easily access our preferred settings by performance type.

    b. Acoustic banners and drapes
       i. All banners and drapes stored
    c. Example choir layout and drape/banner configuration (not included here)

2. Symphony rehearsal
    a. This setting would be appropriate for rehearsals with the same ensembles mentioned above. The purpose of the partially deployed banners is to simulate audience absorption in an unoccupied auditorium.
    b. Acoustic banners and drapes
       i. Upper side wall banners (5 and 6) halfway deployed
       ii. All other drapes and banners stored
    c. Drape/banner configuration (not included)

3. Small ensemble
    a. Typical programs appropriate for this setting include string quartets, piano trios, saxophone quartets, and woodwind quintets.
    b. Acoustic banners and drapes
       i. Upper side wall banners (5 and 6) 100% deployed
       ii. All other drapes and banners stored
    c. Example string quartet and trio musician layouts and drape/banner configuration (not included)

4. Amplified/lecture
    a. Typical programs appropriate for this setting include amplified ensembles, films, lectures, and panel discussions
    b. Acoustic banners and drapes
       i. All drapes and banners 100% deployed
    c. Example big band musician layout and drape/banner configuration (not included)

5. Percussion
    a. Loud acoustic ensembles that require a high degree of clarity would benefit from this setting. Typical ensembles include the University Djembe Ensemble and Percussion Ensemble.
    b. Acoustic banners and drapes
        i. Lower side wall drapes (1 and 2) stored
        ii. All other drapes halfway deployed
    c. Example Djembe Ensemble musician layout and drape/banner configuration (not included)
6. Piano/vocalist
    a. Typical programs appropriate for this setting include solo piano and vocalist performances.
    b. Acoustic banners and drapes (not included)
        i. Upper side wall banners (5 and 6) halfway deployed
        ii. All other drapes and banners stored (identical to symphony rehearsal)
    c. Example musician layouts and drape/banner configuration

Acoustical Measurement Data and Analysis Summary

The maximum RT mid frequency (RT_{mid}) of the auditorium, unoccupied, was measured to be approximately 2.1 seconds, and the minimum RT_{mid} was approximately 1.5 seconds. The adjustable absorptive banners provide 0.6 seconds of variability in the room's response. When the room is fully occupied, it is anticipated that the RTs will be about 0.2 seconds shorter depending on the number of audience members present (Figures A2.11 and A2.12).

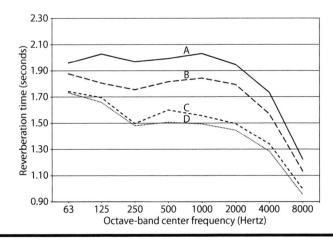

Figure A2.11   Baldwin Auditorium, Duke University, Durham, NC, renovated 2013. (A) Banners 100% stored. (B) Banners 25% deployed. (C) Banners 75% deployed. (D) Banners 100% deployed.

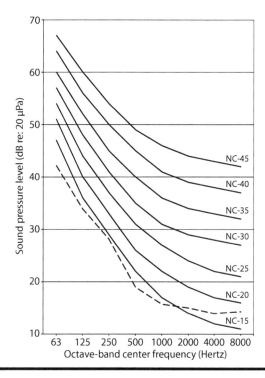

Figure A2.12   Baldwin Auditorium, Duke University, Durham, NC, renovated 2013. NC measurements show NC-15 levels were achieved in the renovation, low enough for professional recording. The perforated wood floor air system is inaudible.

Press

Duke orchestra and chorale doing Beethoven's 9th … with the stage full of musicians the sound and the acoustics are perfect. Can't ever thank you enough.

Tallman Trask III
Executive vice president, Duke University (pers. comm., 2014)

Project Information

DESIGN TEAM

Architect: Pfeiffer Partners Architects
Theater consultant: Theatre Consultants Collaborative

DETAILS

Size: 42,000 SF
685-seat performance hall

BUDGET
$15 million

SCOPE OF WORK
Architectural acoustics
Sound isolation
Mechanical system noise control
Audio/video design

CLIENT
Duke University

COMPLETION DATE
2013

LEED RATING
Pursuing Silver

Appendix III: Benjamin and Marian Schuster Performing Arts Center, Dayton, OH: Three Balconies, Three Lifts, and a Mesh Ceiling

Introduction

The Benjamin and Marian Schuster Performing Arts Center is located in Dayton, Ohio, and was built in 2003 to serve as home to the Dayton Philharmonic, Dayton Opera, and Dayton Ballet as well as popular entertainment and Broadway touring productions. It is owned and operated by the Victoria Theater Association, which also operates the historic 1050-seat Victoria Theater next door. In conjunction with the Schuster's 2300-seat Mead Theater; 150-seat Mathile Theatre; and 17-story office and luxury condo tower, restaurant, and winter garden, these spaces form a thriving downtown performing arts complex (Figure A3.1).

Building Blocks

The Schuster replaced the multi-use 5000-seat Memorial Auditorium, which was built in 1906 to serve as the home for the performing arts in Dayton, as well as sporting events such as boxing and wrestling, graduations, church conventions, and popular music productions. In the 1950s, the flat floor was converted to a very wide, single-balcony proscenium theater with seating for 2000 patrons, an extremely tight stage, and minimal backstage spaces.

One of my first consulting assignments in the 1980s was the acoustic evaluation of Memorial Auditorium's orchestra shell for the Dayton Philharmonic. The philharmonic wanted to know if there was a better shell, one that could improve their sound. I recall politely saying that yes, the old plywood shell could be improved upon, but it would need to break apart into sections small

Figure A3.1 Schuster PAC, Dayton, OH, 2003. The 2300-seat three-balcony hall is considered acoustically superb for orchestral music, opera, and amplified musical events. Note the twinkle lights in the center of the ceiling depicting the night sky.

enough to fit in the elevator for storage in the only space available—the basement. What I wanted to say was that the old hall was a perfect example of an acoustically terrible multi-use hall built in the United States in the 1950s and that a new shell would do little to improve the philharmonic's sound. "Better to build a new hall," I thought and eventually expressed this to them in private.

Challenge

The existing infrastructure did not meet the needs of the client and end users, and significant acoustic modifications would do little to improve the sound. In 1995, Jaffe Holden completed a study that confirmed that the Memorial Hall could not be sufficiently upgraded acoustically, and it was obvious that a new building was needed in order to achieve acoustic success.

Solutions

A number of the acoustic innovations described throughout this book were incorporated into the design of the Schuster Center. These include the incorporation of three stage lifts to allow the orchestra to play forward into the hall, working within the parameters of a multilevel hall design, a sound-transparent ceiling to limit the visual size of the hall without affecting RT or other acoustic criteria, and the inclusion of adjustable acoustic systems to better control acoustic performance for the variety of planned programming (Figures A3.2 through A3.6).

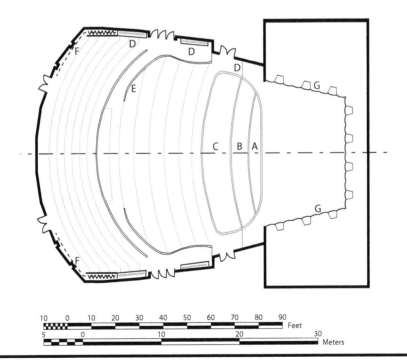

Figure A3.2 Schuster PAC, Dayton, OH, 2003. Orchestra plan. (A–C) Pit lifts/stage extensions. (D) Adjustable acoustic panels on sidewalls. (E) Acoustically shaped parterre wall. (F) Acoustic drapes on rear walls. (G) Orchestra shell towers in deep shell configuration.

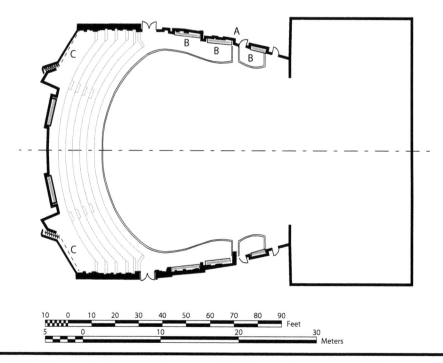

Figure A3.3 **Schuster PAC, Dayton, OH, 2003. First balcony plan. (A) Sound diffusive wall shaping behind. (B) Adjustable acoustic panels that pull up from below. (C) Acoustic drape on rear walls.**

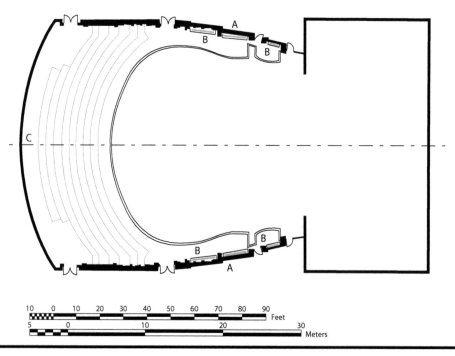

Figure A3.4 **Schuster PAC, Dayton, OH, 2003. Second balcony plan. (A) Sound diffusive wall shaping behind. (B) Adjustable acoustic panels that pull up from below. (C) Acoustic drape on rear walls.**

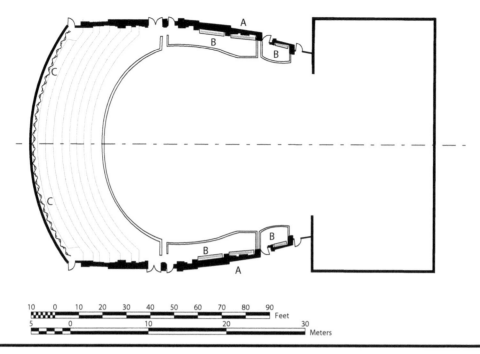

Figure A3.5 Schuster PAC, Dayton, OH, 2003. Third balcony plan. (A) Sound diffusive wall shaping extend to rear wall. (B) Adjustable acoustic panels that pull up from below. (C) Acoustic drape covers entire rear wall.

Creating the Building

Since 1912, the corner of Second and Main Street in downtown Dayton has been a center of the community. A thriving department store at the location closed in 1992. In 1995, a team of community leaders and the county put together a plan for a signature project to create a focal point for downtown's renaissance. In 1999, Jaffe Holden collaborated with Theatre Projects Consultants and Pelli Clarke Pelli Architects to design an outstanding multi-use hall that excels at all types of performance. Construction began in July 2000, and the hall opened in 2003. Featuring a block-long glass-enclosed atrium called the Winter Garden, the center is composed of the 2300-seat Mead Theater, the 150-seat Mathile Theatre, and the Citilites restaurant and bar, as well as an office tower and condominiums.

Architectural Details

Stage Lifts

Large stage extensions, or lifts, are crucial components of the outstanding symphonic acoustics of the Mead. The lifts bring most of the orchestra's string and woodwind sections in front of the proscenium arch and allow full communication with the hall's large acoustic volume. This unique stage is described in detail in Chapter 10. Our acoustic design presented opportunities and challenges that we solved in unique ways—making the Schuster one of the most flexible halls upon which I have worked (Figure A3.7).

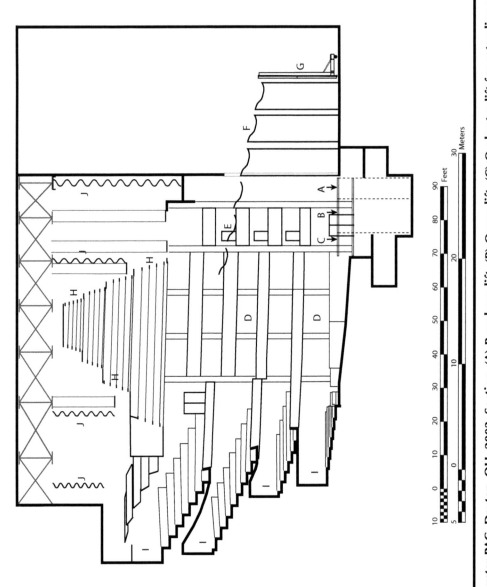

Figure A3.6 Schuster PAC, Dayton, OH, 2003. Section. (A) Broadway lift. (B) Opera lift. (C) Orchestra lift for extending orchestra into the hall. (D) Shaped sidewalls with adjustable acoustic panels. (E) Forestage reflectors store above ceiling. (F) On-stage shell ceilings. (G) Shell towers nest for storage. (H) Sound transparent metal mesh ceiling. (I) Acoustic drapes on rear walls. (J) Acoustic drapes above metal mesh ceiling.

Figure A3.7   Schuster PAC, Dayton, OH, 2003. Dayton Philharmonic in symphonic play position with soloist and most instruments forward of the proscenium.

Multilevel Design

The hall's unique European opera house–style design includes four levels: orchestra, loge, lower balcony, and upper balcony. There are side galleries and one box per side on each level. These multiple levels create acoustically superior shallow balcony overhangs and a visually intimate space where the last seat is only 120 ft. (36.5 m) from the stage. The top tier has a rather steep balcony rake but allows for excellent site lines out to the very edge of the stage extensions.

Forestage reflectors enhance onstage hearing, aid balance, and foster sound projection. Reflectors drop from the attic to tip horizontal at about 35 ft. (10 m) over the orchestra. Throat walls double as part of the acoustic shell to compensate for the orchestra position being located far down stage. Side boxes and portions of the side galleries are left and right of the orchestra in symphony mode creating an experience similar to a vineyard hall.

Adjustable Acoustic Systems

Early designs did not include acoustic wall panels and acoustic drapes on the lower walls. The verticality of the hall meant that all attic acoustic drapes would be positioned very high in the room. Concern over the effectiveness of this strategy led to the decision to position 1 in. thick (25 mm) operable acoustic panels on the side walls and velour drapes on motorized tracks at rear walls on every level. This resulted in a swing of 1.5–2.0 mid-frequency levels of RT.

Sound-Transparent Ceiling

A dramatic ceiling soars over the center of the hall and culminates in a depiction of the night sky as seen by the Wright Brothers during their first flight in 1906. The ceiling is composed of extremely sound-transparent expanded steel mesh on steel frames and a small plaster dome element with fiber optic lights (Figure A3.8).

The architect's original choice of material for the ceiling was perforated aluminum or steel; however, we determined that such materials would inhibit the free path of sound in the upper volume. While most perforated metals are sound-transparent normal to the surface, off axis, there is high-frequency blockage that would reduce M (Mean Free Path) and lower RT at high frequencies. The expanded metal mesh ultimately chosen met the visual criteria for opaqueness when lit from the front and had a higher percentage of open area than the perforated metal. It was also very open to off-axis sound transmission, meaning it caused no degradation of reverberant sound.

The house acoustic volume is approximately 750,000 ft.^{3} (21,000 m^3) and translates to 330 ft.^{3} (9.4 m^3) per audience member. This matched the calculated RT of 2.0 seconds in symphony mode. Interestingly, with the orchestra positioned out on lifts, the seating capacity was reduced from 2325 to about 2100 seats and thereby increased the volume/seat ratio. The hall's dimensions are cube-like at 100 ft. (30.48 m) long, 95 ft. (30.0 m) wide, and 90 ft. (27.4 m) high. All interior surfaces are plaster on block or painted and sealed concrete block in the upper volume in order to provide excellent BR. The throat walls are in the 75–80 ft. (23.0–24.5 m) wide range and are at the limit of recommended dimensions. This means that the hall is very clear and bright for both opera and symphony.

Lifts

The three stage lifts that create the symphony stage extensions are each designed for multiple functions including orchestra pit, seating platform, and stage extension. Lifts also serve as a mechanism to transport seating wagons, which are wheeled wood platforms with the seats permanently affixed.

Figure A3.8   Schuster PAC, Dayton, OH, 2003. View looking up into sound transparent metal mesh ceiling and night sky lighting. Note the forestage reflectors in the ceiling slots in stored position.

When lowered, the first lift is used for Broadway performances and seats up to 35 musicians with an 8-ft. (2.4-m) zone under the stage. This maximizes seat count as only 17 audience chairs are removed.

When lowered, the first and second lifts support an opera orchestra of up to 80 musicians, and the hall capacity is cut by 85 seats (Figure A3.9). These 85 seats are actually on three wagon platforms that store under the seating area.

The third lift only goes from seating level to stage level and is reserved for the use of the symphony. Five seating wagons for 68 seats can be rolled on stage and then stored backstage. Backstage storage is plentiful when the orchestra is positioned so far forward, so there is plenty of storage backstage for the wagons, and stage function is not hindered.

Adjustable Acoustic Systems

Acoustic panels that are 1 in. (2.5 cm) thick pull up from wooden cabinets on three levels to cover throat walls at boxes and at side galleries with mid- and high-frequency sound absorption. These are especially useful for amplified programs because they eliminate throat wall reflections that often reduce speech intelligibility in highly amplified shows (Figure A3.10). Motorized acoustic drapes are also located on rear walls behind seats at all levels. This is not to control RT but rather to control reflections back to the stage during amplified productions.

A careful observer will notice that the orchestra shell ceiling profile was modified from our recommended airfoil shaping by the request of Pelli Clarke Pelli Architects. The collaborative

Figure A3.9 Schuster PAC, Dayton, OH, 2003. Leonard Bernstein's MASS was written immediately after JFK's assassination and performed here. It calls for a large pit orchestra, two choruses plus a boys' choir, a Broadway-sized cast (with a ballet company), marching band, and rock band. Few halls can perform such an immense and breathtaking composition.

Figure A3.10 Schuster PAC, Dayton, OH, 2003. Sidewall, fabric covered fiberglass acoustic panels pull up from a storage cabinet and cover a diffusive surface.

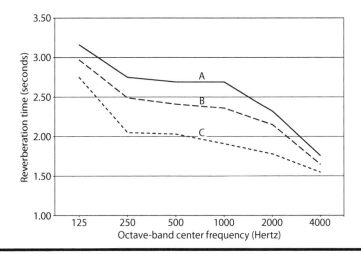

Figure A3.11   Schuster PAC, Dayton, OH, 2003. Unoccupied measurements. (A) Drapes fully stored. (B) Drapes in orchestra play position (the drapes at back of balcony, the proscenium and two catwalk drapes at 600 level deployed, and all others stored). (C) Drapes fully deployed.

solution was to add a slight curve to convex at the leading edge of the shell. This did not create a focusing effect; rather, it was diffusive and a success.

The Schuster Center and the Mead Theater are acoustic and architectural gems tucked away in a modest-sized city in central Ohio.

Measuring Results

Jaffe Holden recorded RT data for a number of hall configurations, as shown in Figures A3.11 and A3.12.

Press

> Architecturally, the Cesar Pelli-designed space is breathtaking. But the real test was whether the sound in Mead Theatre, the orchestra's new home, would live up to its $121 million price tag. By the conclusion of Thursday's concert of Beethoven, Mendelssohn, Vaughan Williams and Stravinsky led by music director Neal Gittleman, it was clear: this is one of the most stunning acoustical spaces in Ohio - perhaps in the nation.
>
> **Janelle Gelfand**
> *The Cincinnati Enquirer (2003)*

Project Information

DESIGN TEAM

Architect: Cesar Pelli and Associates
Theater consultant: Theatre Projects Consultants
Architect of record: GBBN

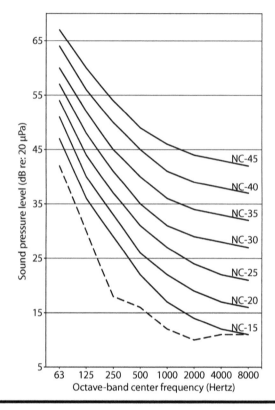

Figure A3.12 Schuster PAC, Dayton, OH, 2003. This hall achieves very low noise levels, less than NC-15. Dotted line indicates measured levels. It is one of the first large halls in the United States with underfloor air supply.

DETAILS

2300-seat hall
Black box theater
Rehearsal space

BUDGET

$77 million

SCOPE OF WORK

Architectural acoustics
Mechanical system noise control
Sound and vibration isolation
Sound reinforcement system

CLIENT

Arts Center Foundation

COMPLETION DATE

2003

Appendix IV: Dallas City Performance Hall, Dallas, TX: Forcing Rough Concrete to Create Soft Sound

Building Blocks

The Dallas City Performance Hall (DCPH) opened in 2012 as the last of the five venues included as part of the Performing Arts Master Plan envisioned to create a cultural destination in Dallas. The other venues in the plan included the Winspear Opera House, the Dee and Charles Wyly Theatre, Annette Strauss Square, and Sammons Performance Park.

The DCPH was conceived to serve the needs of small- and mid-sized arts groups reflecting the diverse communities of Dallas—chamber music, small opera, choral work, ballet, and regional theater (Figure A4.1). The programming process included over 70 arts groups that needed a home that would serve their needs. A multipurpose 750-seat theater with full adjustable acoustic capability, concert enclosure, performance audio and video systems, stage rigging, lighting systems, and an orchestra pit was approved as part of the first phase. A large lobby with event capability and modest backstage support was also approved for a total of 60,000 ft.^{2} (5500 m^2) (Figures A4.2 through A4.5). Phase 1 was paid for through $40 million in city bond funding.

Phase 2 included an additional 64,000 ft.^{2} (6000 m^2) and included two black box theaters and rehearsal spaces but was not funded at the time the main hall opened. Interestingly, right before construction of the DCPH began in 2009, the privately funded AT&T Performing Arts Center Silver parking garage was built under the building—adding complexity and cost to the acoustic isolation design but allowing funding of the motorized orchestra pit lift, which had been eliminated from the project due to budget concerns.

Challenge

An exceedingly varied series of programming was prescribed by the Office of Cultural Affairs. Arts groups desired an intimate jewel box space for performances, whereas the architect imagined

Figure A4.1   Dallas City Performance Hall, Dallas, TX, 2012. A 750-seat multi-use hall in great demand in downtown Dallas. Home to Dallas Symphony concerts, dance, and theatre.

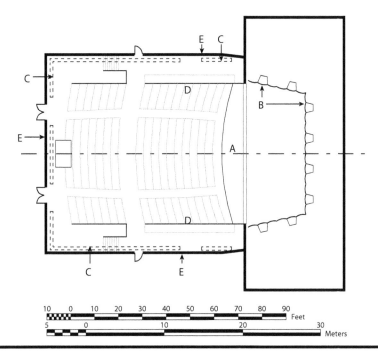

Figure A4.2 Dallas City Performance Hall, Dallas, TX, 2012. Orchestra plan. (A) Pit lift and stage extension. (B) Shell towers store in niche upstage. (C) Acoustic banners suspended from attic above. (D) Portiere wall and rail. (E) Shaped concrete wall for diffusion.

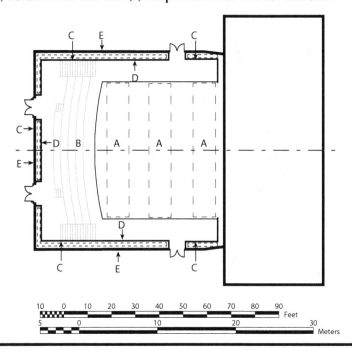

Figure A4.3 Dallas City Performance Hall, Dallas, TX, 2012. Balcony plan. (A) Wood acoustic reflectors above. (B) Balcony level concrete floor. (C) Acoustic banners suspended from above. (D) Openings allow balconies to float off the walls. (E) Concrete walls with diffusive finish.

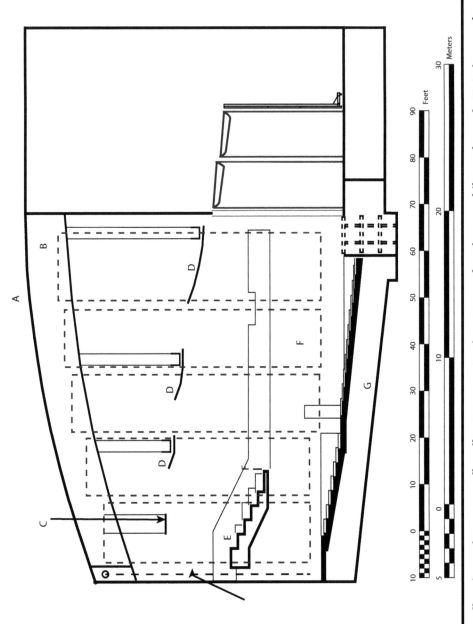

Figure A4.4 Dallas City Performance Hall, Dallas, TX, 2012. Section. (A) Metal roof over multilayer plywood. (B) Attic space for acoustic drape storage. (C) Catwalks exposed in the hall (typ.). (D) Wood acoustic reflectors. (E) Balcony floats off walls. (F) Acoustic banners drop from attic space through slots in the ceiling and past the balconies. (G) Supply air plenum over parking garage (not shown).

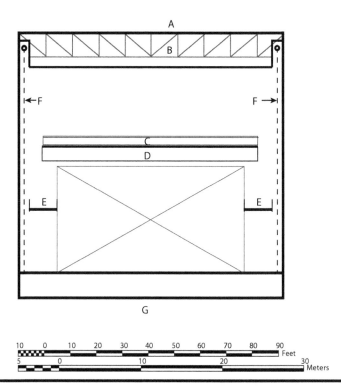

Figure A4.5   Dallas City Performance Hall, Dallas, TX, 2012. Section. (A) Metal roof over multi-layer plywood. (B) Attic space for acoustic drape storage. (C) Catwalks exposed in the hall (typ.). (D) Wood acoustic reflectors. (E) Balcony floats off walls. (F) Acoustic banners drop from attic space through slots in the ceiling and past the balconies. (G) Supply air plenum over parking garage (not shown).

a significant concrete building that could hold its own next to neighboring buildings by Pritzker prize–winning architects.

Solution

We devised a scheme that utilized the advantages of raw concrete for a strong bass ratio (BR) and high sound isolation values and invented a means of removing harshness and acoustic glare with board-formed concrete diffusion furrows on the walls. The desire for a building of significant scale meshed with our desire for a large acoustic hall volume and double-sound barrier roof. White oak acoustic reflector ceilings and oak aisles accentuated by dramatic lighting brought out the sense of warmth and intimacy that the users desired. The result is what Scott Cantrell, music critic for the *Dallas Morning News* (2013), told me is his favorite hall for chamber music performances anywhere.

Creating the Building

The design team included Leigh Breslau of SOM, Corgan Associates architects, Schuler Shook theater consultants, the Dallas Office of Cultural Affairs, the City of Dallas, and Jaffe Holden.

The long design and construction process began in 2004 and continued for eight years, which culminated in an opening night in 2012 that included dozens of performers of community arts groups in the city.

Our acoustic design began with the stage. The room's acoustic design must be secondary; it takes the stage performances' sound and directs, tempers, or augments it for the best possible audience experience.

Acoustic Design Concepts

We determined that for optimal performance, the hall would require a highly adjustable range in reverberation time (RT 60). We planned for adjustability from 1.4 to 2.8 seconds, which is over a 100% range adjustment. That goal was exceeded and resulted in the DCPH being one of the most flexible halls ever built.

In a 750-seat hall, the loudness can be overwhelming, so we reduced the apparent loudness levels (G) so as not to overpower RT and clarity factors. This was done architecturally by raising the ceiling/roof assembly to the height of the stage house, which pumped up the acoustic volume.

Architectural Details

Balcony

The side balconies and rear balcony all float off the walls, creating an air gap that funnels sound down to the seats below it. This was done to eliminate any negative acoustic response that sometimes occurs in seats under the balcony overhang.

Adjustable Acoustic Systems

To tune the hall for the wide range of RTs needed, an innovative acoustic banner system was custom designed using imported English wool serge in double layers. These banners drop out of the attic on motorized cable winches and can be deployed to any location for optimal acoustics (Figure A4.6). We determined acoustic presets for the banners during tuning rehearsals.

Reflectors

The ceiling also has a series of wood acoustic reflectors honed from solid white oak that aid clarity and sound diffusion. They are suspended at the correct height, curve, and angle from the technical catwalk system in order to support the lighting and speaker systems.

Stage Enclosure

In association with Jack Hagler of Schuler Shook, Jaffe Holden designed a flexible wood stage enclosure, or shell, to support and augment soloists, choral ensembles, and full chamber orchestras (Figure A4.7). The shell is actually more suited for sonic blending and onstage hearing rather than projection of sound to the audience. Acoustic excellence begins with the performers; if the musicians can perform at the highest level, then the audience's experience will be exceptional.

Figure A4.6 Dallas City Performance Hall, Dallas, TX, 2012. Acoustic banners made of two layers of wool serge can drop down to cover the side and rear walls. This view shows the drapes fully extended.

Noise Control

A challenge of this site was the constant noise from the nearby airport and the busy parking garage below the hall. Sound barrier materials were incorporated that would completely shield the hall from unwanted outside sounds.

The air systems needed to be both energy efficient and extremely quiet. Our measurements confirmed that air noise is inaudible (below NC-15) and close to the threshold of hearing in all areas of the hall, a remarkable feat given the modest budget and noisy site.

Figure A4.7 Dallas City Performance Hall, Dallas, TX, 2012. The tunable orchestra shell has a light wood veneer to match the interior's white oak. Dramatic lighting is used by the Dallas Symphony for a concert of Copland and Reich in 2015.

Concrete Walls

A strong BR of 1.25 or more was desired. This equates to a rich, resonate bass response, and very high clarity (C80) levels. We advised the architects to use complex sound diffusion on a wall of highly finished concrete surfaces poured in place to mellow and blend large music ensembles. The architects used board-formed concrete with a rough wood board texture to temper acoustic glare and articulated the board form in and out in horizontal strips to our specifications. This would enhance the acoustic response of the parallel side walls as well as diffuse and blend the lateral sound reflections.

Measuring Results

The Jaffe Holden team copiously measured the hall's acoustics at many banner settings to find the optimal settings for rehearsals and performances. With all banners stored in the attic and with the shell on stage, we exceeded 3.5 seconds of RT in the unoccupied space. It is quite a comparison to the 1.75 seconds where all banners were deployed. It dropped further with an audience present and the shell removed—an estimated 1.4 seconds, all without harming the sound's timbre or excellent sonic quality (Figures A4.8 and A4.9).

Acoustic Tuning Notes 9/16 and 9/17, 2012

The following excerpt is from my notes on the acoustic tuning in the fall of 2012. They describe some of my thinking process and conclusions during those two exciting days.

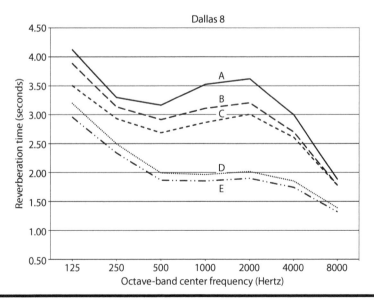

Figure A4.8 Dallas City Performance Hall, Dallas, TX, 2012. RT measurements in unoccupied hall with shell. (A) Banners fully stored. (B) Side banners only deployed. (C) Rear banners only deployed. (D) Banners 6/8 only deployed. (E) Banners fully deployed.

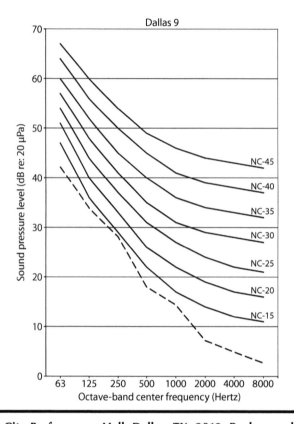

Figure A4.9 Dallas City Performance Hall, Dallas, TX, 2012. Background noise NC measurements (dashed line). Hall meets quietest criteria of NC-15 with all systems running.

I arrived at the hall for tuning and commissioning to find the orchestra shell set with preliminary angles that looked too flat and all acoustic drapes deployed. This setting would render the hall quite dry and lifeless, which, even at this setting, frankly sounded better than most halls.

The pianist for this day was Jerry Stephens, the accompanist for the Dallas Symphony Chorus. Jerry was easy to work with and immediately began a discussion with me about the goals and the plan for his 3-hour tuning session. I explained the concept of tuning as a movement from out of tune to in tune, much as a piano tuner forces a note out and brings it back into alignment.

The place to start was with all banners deployed and the piano slightly upstage of the fire curtain line. In this location, the piano sounded flat and dull on the main floor but live and full in the balcony. The drapes were doing their job of dulling the sound of the reverberation and dampening early lateral sound reflections off the side walls. This was expected and reassuring, but the effectiveness of the banners in the lower positions was surprising.

The second day of open-house events was more acoustically focused. Today, we enlisted the help of the Orchestra of New Spain, the Lone Star Wind Orchestra, and the Women's Chorus of Dallas.

We have yet to set the presets for the 13 acoustic banners other than all exposed, all stored, and choral. A preliminary setting of 40% deployed on the sidewalls with 100% rear drapes deployed, and another one with 60% walls deployed was set.

We are also checking the shell angles, and I believe that the front ceiling piece is set too flat for correct reflections from the conductor and concertmaster and will need to be tipped to a more aggressive angle. Carlos Rivera checked the angle using ray tracing and AutoCAD. Thank goodness he is knowledgeable with these programs—I would be doing it with tracing paper and mirrors. We determined that the angle needed to be steeper in order to send stronger early reflections from the strings to the center of the orchestra seats. This creates a better "bite" to the string sound where the attack of the bow on the string is palpable and brilliant.

Results of the tuning so far are gratifying and thrilling, greatly aided by the excellent staff of Russell Dyer, general manager, and Rob Crain, technical director.

Press

> Dallas City Performance Hall proved to be an acoustic marvel. All that wood on the seats, the floor and the overhang created a bouncing effect perfect for the sounds instruments make. Basses boomed without turning fuzzy, drums snapped without pounding and guitars delivered crisp notes that never distorted.

> **Mario Tarradell**
> *Music critic, Dallas Morning News (2012)*

Project Information

DESIGN TEAM

Architect: Skidmore, Owings & Merrill, LLP
Architect of record: Corgan Associates, Inc.
Theater consultant: Schuler Shook

DETAILS

60,000 SF
750-seat proscenium theater
Backstage support areas

SCOPE OF WORK

Architectural acoustics
Mechanical system noise control
Sound isolation
Audio and visual design

CLIENT

City of Dallas

COMPLETION DATE

2013

LEED RATING

Pursuing LEED Silver

Appendix V: Daegu Opera House, Daegu, South Korea: Korea's First Multi-Use Theater

Building Blocks

The city of Daegu, formerly known as Taegu, wanted to create a downtown cultural district and entertainment venue to serve a large variety of performances. This would require the relocation of the downtown industrial factories to the outskirts of the city. Cheil Industries, the parent company of Samsung Electronics, agreed to relocate their facilities and in return pledged to gift the city with a 1500-seat multi-use hall. While named the Daegu Opera House, the programming was not limited to only opera. The space served a wide variety of performances including Broadway productions, dance, headliner concerts, and small classical ensembles (Figures A5.1 and A5.2). The Daegu Opera House opened in 2004 and was designed by Samoo Architects, Fisher Dachs Associates (FDA), and Jaffe Holden. The space was acclaimed for its unusual visual and aural intimacy as well as the first large-scale use of adjustable acoustic systems in Korea. The only Korean International Opera Festival is held here annually in the fall.

The use of adjustable acoustics was quite foreign to the technical reviewers from Samsung and was not immediately embraced. Acoustic modeling and case studies of similar projects in the United States convinced the client of its value. Adjustable acoustics were implemented with the use of manually operated wall systems, which provided simplicity and a low initial cost. A complex series of 48 motorized acoustic banners in concentric rings would drop down from the dome area. Also, a remarkably flexible double orchestra pit complements the adjustable acoustic systems.

Challenges

Many of the standardly employed adjustable acoustic systems were unfamiliar to the client.

Figure A5.1 Daegu Opera House, Daegu, South Korea, 2004. The roof profile is reminiscent of a piano body to hide the stage house tower.

Figure A5.2 Daegu Opera House, Daegu, South Korea, 2004. The 1500-seat hall hosts a wide range of performances and is the first to use adjustable acoustic systems in Korea.

Solutions

Sophisticated dome banners, side wall drapes, and multiple pit levels created very flexible acoustics for the first time in Korea.

Creating the Building

Design Process

Early in the design, a traditional horseshoe shape was suggested by Josh Dachs of FDA. The space would have three balcony levels that would wrap around the hall's side walls with side galleries. The acoustics of horseshoe opera house multi-use halls vary from a more traditional hall design. Achieving strong and rich reverberation levels in a hall of this shape is a challenge because the audience covers the side and rear walls thus restricting the development of side-to-side reverberant buildup. We agreed to the design direction and defined an optimal reverberation time (RT) range of 1.2–1.6 seconds given the 1500-seat capacity and wide program of events. Our initial reaction was to limit the overhangs of the balcony (as described

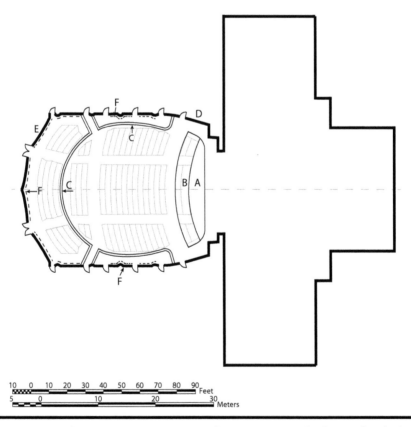

Figure A5.3   Daegu Opera House, Daegu, South Korea, 2004. Orchestra level plan. (A) Pit lift 1/ stage extension. (B) Pit lift 2/stage extension. (C) Parterre wall. (D) Throat wall angled at about 7°. (E) Rear wall convex shaping. (F) Acoustic drapes on walls (manual).

in Chapter 11) especially on the main floor to five rows maximum. Also, we recommended to minimize the side gallery depths to prevent trapping the sound. The volume defined by the walls above the galleries supports the 1.6-second reverberation, and the RT is effectively controlled when the banners drop into the volume restricting the mean free path (Figures A5.3 through A5.6).

Architectural Details

Pit

In a multi-use hall, the forestage area and the pit are major acoustic challenges. This is described fully in Chapter 10. Our concept was to fix the forestage ceiling rather than provide a movable forestage element. This concept was implemented to reduce construction costs and to simplify the acoustic and theatrical design. We located the ceiling at a height that supports musicians in the pit with overhead reflections. The sound reflects from the performers into the orchestra-level seats where balances with pit orchestras tend to be most problematic. Classical music and symphonic works were not part of the program, so an orchestra shell was eliminated.

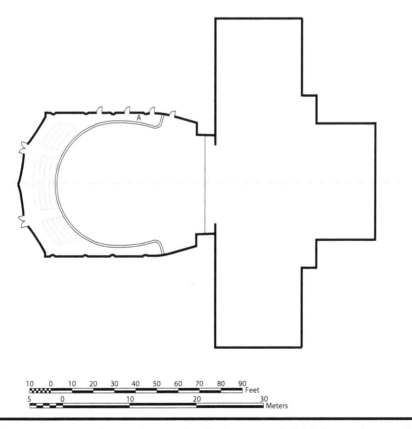

Figure A5.4   Daegu Opera House, Daegu, South Korea, 2004. 1st Balcony plan. (A) Adjustable acoustic drapes on sidewalls (manual) over shaped sidewalls for diffusion.

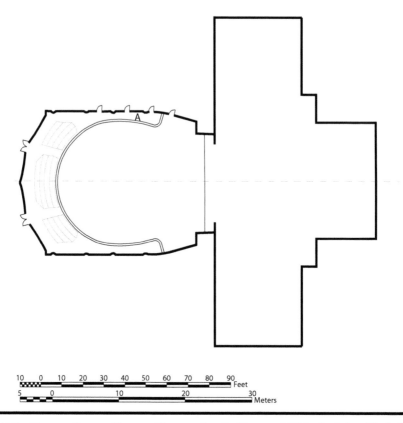

Figure A5.5 Daegu Opera House, Daegu, South Korea, 2004. 2nd to 4th balcony plan. (A) Adjustable acoustic drapes on sidewalls (manual) over shaped sidewalls for diffusion.

Volume

The volume is set at a reduced level to achieve the RT of 1.6 seconds when all of the adjustable acoustic systems are stored. This RT might seem excessive when considering that the RT of European powerhouses is well below that figure. The Teatro alla Scala in Milan, Italy, is 1.24 seconds, and the Wiener Staatsoper in Vienna, Austria, is 1.36 seconds. However, American opera houses such as the Kennedy Center Opera House and the Wortham Center in Houston are longer at 1.6–1.7 seconds mid frequency. American opera houses are much larger in capacity, typically over 2000 seats, which leads to a large acoustic volume and results in a naturally longer RT. Our goal was to define the volume so that no permanent sound absorption would be needed to reach the 1.6 seconds target.

Dome Ceiling

Initially, the ceiling was proposed to be a concave and a sound-focusing dome shape. We suggested refinements to diffuse and flatten the dome in the center portion. For example, we recommended to use tight radius coves to give a dome appearance without acoustic focusing. The dome was also an ideal location for the motorized acoustic banners. At the rear of the dome, FDA desired a major lighting position. We opened the dome's curve and provided a theatrical lighting position, eliminating a possible late-arriving reflection surface. Since the hall's program did not involve

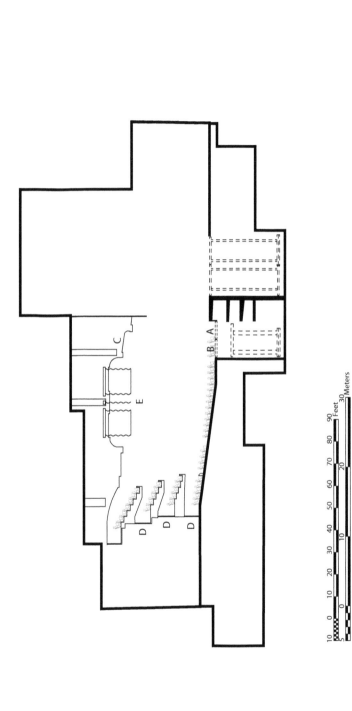

Figure A5.6 Daegu Opera House, Daegu, South Korea, 2004. Section. (A) Pit lift 1/stage extension. (B) Pit lift 2/stage extension. (C) Ceiling over pit area shaped for reflections back to pit musicians. (D and E) Adjustable acoustic banners.

Figure A5.7 Daegu Opera House, Daegu, South Korea, 2004. The 1500-seat hall hosts a western opera festival each year, but can present all manner of performances. Acoustic banners drop from the domed ceiling to control RT.

symphonic performances, we rejected locating adjustable acoustic systems in the attic volume and designed a closed ceiling. The adjustable acoustic systems are a hybrid of 48 motorized roll-down acoustic banners in two concentric rings that drop from within the dome. There are also additional manual drapes on the side walls at all levels (Figure A5.7).

Wall Shaping

The width of the hall was a narrow 82 ft. (25 m); as a result, the throat walls could flare naturally from the proscenium area at 7°. Larger halls such as the Long Center or Wagner Noël require a step design to achieve this flare. We simplified the shape of the orchestra plan to be rectilinear overall while inserting a horseshoe orchestra level for earlier lateral reflections and C80 enhancement to the center orchestra seat. The balcony and gallery front created the horseshoe shape on the upper levels desired by FDA and Samoo, while the shallow soffits under the side galleries were ideal for second-order reflections from the walls and soffit plaster surfaces.

In order to control RT and early and lateral reflections, velour drapes were recessed in shallow pockets along all the throat and side walls. The drapes can easily be hand drawn on each of the three levels. If the design had the galleries step down toward the stage, the drape configuration would have been very complex and perhaps unworkable. The rear/lobby walls were desired to follow the concave seating curve, and we reshaped them into three convex, nonfocusing diffusive elements. Additional diffusion surfaces include the side wall acoustic banner enclosures, convex balcony fronts, and the tight-radius curved-ceiling elements.

Pit Details

The Daegu Opera House orchestra pit is sophisticated and complex with multiple configurations and elevations possible (Figure A5.8). It is an excellent example of our acoustic design described in Chapter 10. The double lifts and seating areas under the cantilevered stage allow for the capacity to range from 20 Broadway musicians to over 100 musicians for the grand opera. The pit lifts descend to lower basement levels, allowing for convenient scenic loading. The lifts can be fully elevated to the stage level for concerts on the forestage and have seating wagons that allow the audience to be seated on either lift.

Measuring Results

Acoustic measurements have not been released by the client, but one was reported to be 1.2–1.6 seconds RT. Reports from the users and audiences are that the hall's acoustics are remarkably clear, intimate, and articulate. This is a result of the architect following the acoustic design concepts for the pit design, early reflection surfaces (throat walls at 7°), diffusion and shaping criteria, lateral reflections from soffits, and correct acoustic volumes. The underfloor air supply systems have an extremely low background noise level, and the poured-in-place concrete walls and roof inherently block outside sound transmission.

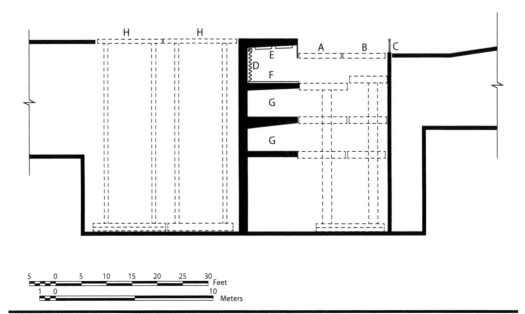

Figure A5.8   Daegu Opera House, Daegu, South Korea, 2004. Orchestra pit design allows for high level of flexibility. (A) Small pit lift for Broadway style productions. (B) Opera pit expands area to allow up to 100 musicians. (C) Solid pit rail. (D) Velour drape on upstage wall. (E) Acoustic panels on underside of stage structure. (F) Wood floor similar to stage floor. (G) Storage areas for seat wagons. (H) Stage lifts allow large sets to be stored on lower levels without disassembling them.

Press

The facility guide for the Daegu Opera House highlights the following achievements:

> The Daegu Opera House has made efforts to implement perfect acoustics as an exclusive theater for opera. It satisfies acoustically appropriate hall volume (W:25m × L:34m × H:15.48m) and auditorium size so that acoustic demands can be satisfied from the rear auditorium and side balcony seats. In order to make optimal acoustic effect, the reverberation time was designed to be 1.2 to about 1.6 seconds with only 0.02 seconds difference between direct sound and indirect sound; also there is no echo (resonance) detected.

Daegu Opera House Foundation
http://english.daeguoperahouse.org/ (n.d.)

Project Information

DESIGN TEAM

Architect: Samoo Architects & Engineers
Theater consultant: Fisher Dachs Associates

DETAILS
185,000 SF (17,000 m^2)
1508-seat hall

BUDGET
44 billion won

SCOPE OF WORK
Architectural acoustics
Mechanical system noise control
Sound and vibration isolation
Sound reinforcement system
AV system design

CLIENT
Cheil Industries

COMPLETION DATE
2003

Appendix VI: Grand Hall, Lee Shau Kee Lecture Centre at Hong Kong University, Hong Kong: Transforming a Lecture Theater into a Grand Hall

Building Blocks

My involvement in the Grand Hall began in 2008 with a phone call from Fred Shen of Shen Milsom & Wilke (SMW), an AV and acoustic consulting firm with offices worldwide. Fred's firm had been awarded the acoustic and AV design for the new Centennial Campus at the prestigious Hong Kong University (HKU), and he asked me to join the team as experts in multi-use hall design in association with the local architect group Wong & Ouyang. The project would involve creating a multi-use 1000-seat hall in the lower level of a new academic building carved out of a mountainside. It sounded fantastic.

Challenges

The planning of the building was well underway, and the lecture hall had been all but designed. Our challenge was to redesign it into a multi-use hall for chamber music, full orchestra, and amplified entertainment and for use as a lecture hall for classes and scholarly presentations. Schematic design was completed, so the basic shape, volume, and dimensions of the facilities were set in stone. Additional challenges included noise from a pedestrian main street and outdoor plaza on top of the hall and water bubbling up through volcanic rock below the hall. Every meter of further excavation would cost 1 million Hong Kong dollars.

Solutions

Cosmopolitan Hong Kong surprisingly lacks acoustically excellent recital and concert halls. The university desired that the Grand Hall be acoustically enhanced to be suitable for an orchestra of up to 100 members and for the hall to have the rich, full acoustics of a concert hall. This meant that the hall needed a rich bass ratio (BR), strong clarity and envelopment, a very quiet NC-15 background level, and an adjustable reverberation time (RT) of 1.2–1.8 seconds in order to accommodate amplified shows and lectures. This was no small task.

We overlaid the hall's plans with that of Alice Tully Hall at Lincoln Center, under construction at the time in New York. The Alice Tully Hall case study (Appendix I) describes our efforts on the project. Tully was a renovation, but the new hall at HKU was similar in plan, section, and seat count. It, too, lacked a stage house, was somewhat wide and low, and had a steep amphitheater seating rise with a modest balcony. We drew from the unfinished Alice Tully Hall as the acoustic design direction for the new hall in Hong Kong (Figure A6.1).

Creating the Building

Design Process

After meeting with the university's leadership, it was determined that the Grand Hall's program was to include classical concerts, music ensembles, amplified music, conferences lectures, video presentations, and other conference activities. Events that were not supported included dance, opera, musical theater, film presentations, or activities that demanded rigging, theatrical sets, or an orchestra pit.

The concert events could include soloists, chamber music, amplified jazz and contemporary music, and a full orchestra with 100+ musicians and chorus. Programming would not include rock 'n' roll or other highly amplified events but by the time the opening night arrived, the events included amplified rock bands and a film presentation.

The acoustic quality for audiences and musicians when presenting classical concerts is expected to be of the highest level, equal to a fine concert hall for local and touring orchestras. Likewise, high levels of speech intelligibility should be delivered to all 1000 audience members for speech events (Figures A6.2 through A6.4).

Architectural Details

Stage Acoustics

Creating proper stage acoustics for orchestral music is critical for delivering high-quality sound to the audience. The stage area including the walls, floor, and ceiling of a multi-use hall must do a number of things very well, including projecting sound to the audience in a well-balanced and blended way. It must provide a blending chamber for the conductor to balance instruments with a wide range of power levels, from violin to brass. Finally, the stage area must provide surfaces to optimize onstage hearing and communications between the musicians (Figure A6.5).

The preliminary drawings showed a stage area that is slightly undersized for a full symphony and with little room for soloists and a chorus. We recommended that it be 50 ft. (15.24 m) wide

Figure A6.1   Grand Hall, University of Hong Kong, China, 2013. This 1000-seat multi-use hall has the first acoustic banners system in China. It has garnered excellent reviews for both symphonic acoustics and silent air systems.

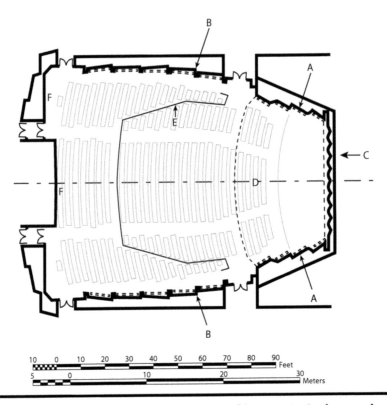

Figure A6.2 Grand Hall, University of Hong Kong, China, 2013. Orchestra plan. (A) Wood stage walls shaped for blending and projection. (B) Stepped sidewalls and columns provide clarity and diffusion. (C) Upstage wall is diffusive and can be covered with a horizontal drape. (D) Stage extension used for larger ensembles. (E) Partier wall. (F) Rear wall perpendicular to center line.

(the average width, with 58 ft. at the stage edge) and 45 ft. (13.7 m) deep, with an extension into the audience of 12 ft. (3.67 m) when a full chorus of 100–150 voices is present.

Stage walls (left, right, and rear) needed to be shaped to project sound and provide blending. Wood walls of modest mass, shaped in chevrons or cylinders, would need to be developed. There was no theater planner on the project, so we suggested that the wing space be provided stage right and left and the use of orchestral shell towers or pivoting doors be promoted to allow access for instruments, artists, and equipment.

The stage ceiling shown in the study and the early drawings included an ill-advised dropped proscenium. This element would trap sound behind it and not provide an even projection of sound. Musicians seated behind it would have a vastly different playing experience to those seated in front of it. Balance would also be difficult to achieve. This element was, therefore, eliminated.

We advocated that the ceiling (clouds) be shaped and angled to blend and project the sound of classic ensembles and be adjustable, or tunable, in order to provide optimal sound. Blending, balance, and projection are interdependent variables. We suggested that they should range from 25 to 30 ft. (7.6–9.1 m) off the stage and be able to be angled and located at various levels.

Adjustable acoustics were not part of the early drawings, so I advocated for acoustically absorptive banner and drapes. These would need to be exposed along the rear wall of the stage and the hall to dampen sound and reverberation of amplified events and speech and then be retracted into

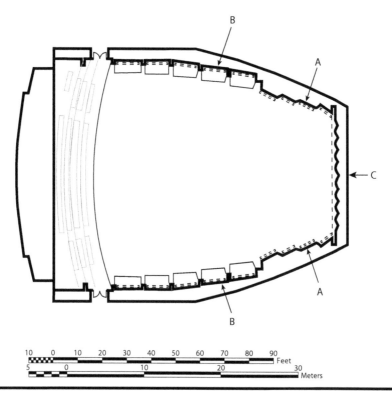

Figure A6.3 Grand Hall, University of Hong Kong, China, 2013. Balcony plan. (A) Shaped stage walls also have vertically deployed acoustic banners. (B) 10 sidewall acoustic banners cover upper surfaces above a horizontal clarity reflector. (C) Horizontal tracking motorized drape on upstage wall stores in pockets off stage.

pockets for classical events. We suggested that the ceiling clouds be designed to be rotated vertically to allow for more complex rigging and lighting, as was planned for Alice Tully Hall.

Hall Acoustics

One of the most critical acoustic criteria is the creation of adequate acoustic volume. Proper volume allows for the generation and sustainment of reverberation. It allows the room to breathe and to absorb high sound levels for the stage without overloading or oversaturation, which leads to a harsh, distorted sound.

Early drawings indicated a volume of 7.8 m³ (275 ft.³) per person. Our experience shows that for a hall of 1000 seats, the volume should be closer to between 10 and 11 m³ (350–390 ft.³). If clarity and brightness can be provided at the same time, both a warm, rich reverberation and a clear, intimate sound can exist that does not overload a full orchestra and chorus. Chapter 8 discusses volume strategies in more detail.

Another early suggestion put forth by the university was to include adjustable acoustic volume, one that used moving ceiling elements to adjust volume and control RT. This technique was widely used in the United States throughout the '70s and '80s but was proved impractical and costly due to achieving only modest acoustic results. The team decided not to move forward in this direction.

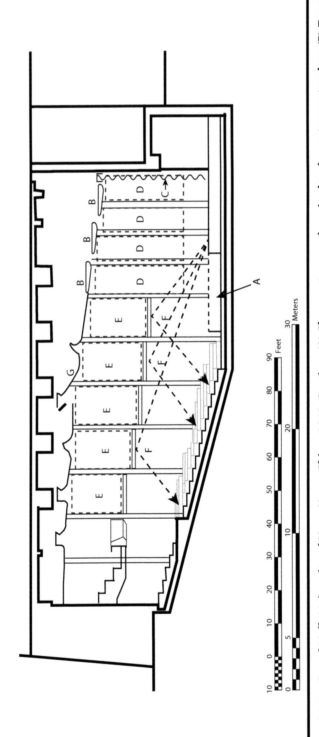

Figure A6.4 Grand Hall, University of Hong Kong, China, 2013. Section. (A) Floor system can be raised to form stage extension. **(B)** Tunable airfoil shaped reflectors over the stage and stage extension. **(C)** Upstage wall and full height acoustic drape. **(D)** 8-stage dual-layer acoustic banners can extend to floor level. **(E)** 10 dual-layer acoustic banners can cover upper walls. **(F)** Sidewall acoustic reflector hides lighting. Arrows show enhanced clarity reflections. **(G)** Shaped ceiling provides diffusion and reflections and is hung below isolation ceiling. Street level Public Plaza is over the hall.

Figure A6.5 Grand Hall, University of Hong Kong, China, 2013. Tuning concerts were used to set ceiling reflector angles and acoustic banners.

Working in China on a complex project of this nature required me to develop innovative new designs for acoustics, sound isolation, and noiseless HVAC systems, many never before used in this part of the world. The room acoustic design solution was similar to what I used at Alice Tully Hall. That case study includes a discussion about room acoustic shaping for RT, C80, diffusion, BR, and so on. This case study, on the other hand, will focus on details for designing the Grand Hall within the lower levels of a noisy campus center building adjacent to the large chiller plant, parking garage, and multiple escalators.

Adjustable Acoustic Systems

An acoustic banner system to control RT was a new concept for China. I had used similar systems at Alice Tully Hall, Harris Hall in Chicago, Legacy Hall at Columbus State University, and other locations in the United States, but our preferred supplier from the United States was deemed to be too costly. We advocated for acoustic banners to add absorption into the room, absorb loudness and energy, and reduce RT and reflections. Chapter 15 includes an in-depth discussion on banner details. In the end, I used a common motorized sun shade system with wireless remote controllers, an equipment that is both reliable and effective. The acoustic banners, made of wool serge instead of shade cloth, descend from enclosures to cover the side walls in a color that matched the wood veneers (Figure A6.6).

A local sun shade manufacturer in Hong Kong won the bid and submitted shop drawings. A trip to the manufacturer's shop was scheduled so that we could see the mock-up and ensure that

Figure A6.6 Grand Hall, University of Hong Kong, China, 2013. Acoustic banners are deployed both on stage and in the hall for amplified programs. All banners are deploying during this action photo. Stage extension is at the lowest level and seated.

they could build and install machines to successfully move double layers of wool serge. The owner's son was a bright young engineer trained in Calgary who spoke excellent English. He explained how the company had modified the standard window sun shades with double layers of wool serge acoustic banner fabric and added a clever wireless remote control to operate them from anywhere in the theater. That innovation made the process of acoustic tuning a breeze.

Noise Control

The Grand Hall's location on the lowest level of an active classroom building presented major sound isolation and HVAC noise control issues. A campus main street was planned for the top of the hall and would connect many buildings set in the mountainside. This street would be very active with foot traffic, rolling carts, street fairs, and concerts. It would be fully exposed to the outdoors. Surrounding the hall on the left and rear would be conference spaces, classrooms, music department practice rooms, and rehearsal halls. To the right was the main chiller plant with four very large chiller machines the size of locomotives, pumps, and electrical gear. Escalators and elevators surround the hall on all sides.

The need was established to have a very quiet room in order to be able to hear the nuances of the sound reflections and to hear the full energy of the RT. Actually, it was determined that this would be the quietest hall in Hong Kong. If the noise floor was not very low, the steep amphi-theater seating rake would rapidly reduce the G level and the energy of the RT, making the hall appear to be less live. The concept for the box-in-box sound isolation system was rather straightfor-ward, but the execution was excruciatingly complex. It demanded construction techniques never before used in a building of this type in Hong Kong.

Box in Box and NC-15

The concept for the box-in-box was straightforward. Massive concrete walls and slabs would make up the outer box structure. A three-layer drywall floating inner box ceiling was suspended below the ceiling slab on spring isolators with batt insulation in the void from which the decorative ceil-ing was hung. We designed a three-layer drywall floating inner box walls inside the mass walls with batt insulation behind, and neoprene braces acted as a resilient brace to the outer box. For the seating and stage area, we floated a 4 in. (10 cm) thick inner box concrete floor slab on neoprene pads on top of the structural floor.

This plan did not go smoothly. The construction was much more complex because the structure was set and could not take the load of the floating slab in the balcony. The mechanical engineers wanted to supply air through the floor, potentially drilling hundreds of sound leak holes in the box-in-box. Plus, there was no contractor with the experience to build such a complex isolation system in Hong Kong or even in China. The solution came after months of discussion with engineers and architects, all guided by SMW's project manager and his knowledge of China's unique construc-tion techniques. He arranged for the floating constructions to be engineered and installed by the vibration isolation supplier and contractor (Kinetics) who had local experience with sound studios and installations in China. We never could have achieved the NC-15 result without this expertise.

The floating concrete floor in the balcony was changed to a timber sandwich system of plywood and drywall. This achieved a modest but acceptable isolation performance because the weight of the desired floating concrete would have caused a collapse.

More Details

Sound and vibration isolation systems are not the subject of this appendix; however, the reader might be interested in how this was achieved in China. The floating side walls could rest on the floating concrete slab, but the floating walls were required to be structurally framed in massive and very heavy steel framing members. This would ensure that the finished walls, supplied by a fit-out contractor, would have a solid structure from which to work. Some of these inner box walls were almost 50 ft. (15.2 m) tall, and each piece of steel structure was cut down to fit in an elevator to enter the hall.

The engineers wanted to supply air at each of the 1200 seats. The challenge was how to get ducts to the seating without destroying the box-in-box floating concrete floor. The solution was a unique and ingenious floor air system that, against all odds, achieved NC-15. The ducted system fed a continuous perforated air supply grille on the vertical riser of each seating step in the hall. The duct sat on top of the floating concrete inner box floor, and the seating floor structure was built over and around the ducts using steel angles, thick steel plates spanning the angles, and a plywood top decking onto which the seats were bolted.

Mock-ups were critical. Kinetics, the vibration isolation manufacturer, made mock-ups of the walls, ceiling, and floor assemblies for our review. The HVAC contractor built a section of the floor ducts and grille system and tested it in an anechoic chamber under actual airflow conditions to prove that the designed airflow could be delivered at NC-15. It succeeded, and we had a plan.

Measuring Results

I was thrilled with the sound of the hall from the very first rehearsal. I was very pleased with the bright, clear sound that leapt off the stage. Musicians had excellent room response. The sound in the hall is modestly immersive but has a bell-like clarity with warm bass and a rich sonority.

The RT and BR are greater than Alice Tully Hall in its maximum mode because of the higher volume and more massive side wall and ceiling construction. The upstage acoustic drapes and the larger surface area of the hall's side wall acoustic banners provided a greater swing in RT than Tully as well. Measurements confirmed that the NC-15 background noise levels were achieved with ease. Measurements showed levels 5–10 dB quieter at many frequencies (Figures A6.7 and A6.8).

Hall Tuning

Hall tuning occurred from December 2–9, 2012.

My general conclusion was that the hall should be used with the large stage extension as the normal condition for all acoustic performances.

In terms of banners, it was determined that all acoustic performances benefitted from banners being stored. During acoustic rehearsals, the 10 house drapes could be deployed to better simulate the effect of the audience. Banners should be deployed not for unamplified performance but rather for rehearsal mode and amplified productions.

The upstage drape was only to be used for amplified productions, lectures, and film events.

In terms of ceiling reflectors, we determined the presets and that reflectors could remain in position for all events. Reflector heights off stage should be 9.8, 8.8, and 8.1 m (32.1, 28.9, and 26.6 ft.) to the bottom of the reflectors. Ceiling reflector angles from upstage to downstage should be 13.4°, 11.2°, and 11.2°.

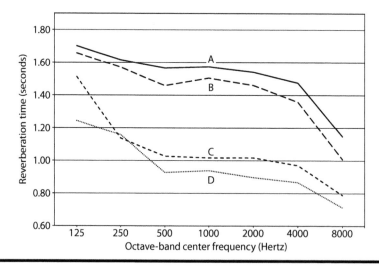

Figure A6.7 Grand Hall, University of Hong Kong, China, 2013. Both occupied and unoccupied RT measurements taken with various settings. (A) Unoccupied, all banners stored. (B) Occupied, all banners stored, stage extension in place. (C) Unoccupied, all banners deployed. (D) Occupied, all banners deployed, no stage extension.

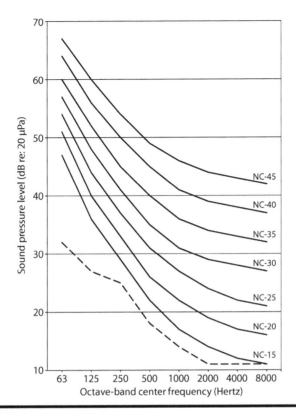

Figure A6.8   Grand Hall, University of Hong Kong, China, 2013. HVAC and ambient noise levels are low, below NC-15, due to exceptionally fine construction (dashed line) and reported to be the quietest hall in Hong Kong.

Press

Reviews from the users and press have been overwhelmingly positive and supportive.

> We had the Hong Kong Philharmonic give a concert on Sunday in the new HKU venue—74 players plus violin soloist, playing Brahms, Ravel, Chausson, and Berlioz. Far from being loud, dry and overwhelming (our worst fears!) the music sounded absolutely wonderful, almost like a CD (this is meant to be a compliment!). Clear detailed sound, well-blended, with a touch of warmth and all at an appropriate volume. Needless to say, the event was a great success for the hall. It's the best acoustics in HK.

> **Professor Daniel K. L. Chua**
> *Head of the School of Humanities, Professor of music,*
> *University of Hong Kong (pers. comm., 2013)*

Project Information

DESIGN TEAM

Architect: Wong & Ouyang Architects
Acoustician: Jaffe Holden in association with SMW

DETAILS

Size: 1000-seat theater

SCOPE OF WORK

Architectural acoustics
Building systems noise and vibration control
Sound and vibration isolation
Adjustable acoustics
Concert enclosure

CLIENT

Hong Kong University

COMPLETION DATE

2012

LEED RATING

Pursuing Platinum LEED

Grand Hall at HKU Case Study

I prepared these notes when in Hong Kong for the tuning in late 2012. These notes describe my thoughts and feelings during the 5 day tuning process and might be of interest to the reader.

Notes on Tuning

Big Ideas, December 2–9, 2012

The hall should be used with the portable stage extension, extending the stage forward as far as possible, as the normal condition for all acoustic performances. This adds a nice bloom to the sound, with reverberation time (RT) at about 1.5 seconds.

All acoustic performances benefit from the banners that are being stored and should not be deployed in any unamplified performance. The banners should be used for rehearsal mode and amplified productions. The upstage drape is only to be used for amplified productions, lectures, and film events. This drops the RT to about 1 second.

For acoustic rehearsals, the 10 house drapes may be deployed to better simulate the effect of the audience. They should be stored for performances.

The ceiling reflectors are preset and can remain in position for all events. The heights offstage are 9.8, 8.8, and 8.1 m to the bottom of the reflectors. The ceiling reflector angles are (from upstage to downstage) 13.4°, 11.2°, and 11.2°.

Tuning Notes, December 2, 2012

Choral ensemble (about 20 student singers, 1 choral conductor, and 1 piano accompanist)

1. All the drapes were extended, and the ceilings were set at nominal conditions. (See the "Big Ideas, December 2–9, 2012" section for the elevations and angles.)

2. The upstage drape made onstage hearing difficult and muffled the singers, so it was quickly stored.

3. The musicians were moved downstage 3 m (9.8 ft.) away from the rear woodwall, but this was not liked by the conductor or the musicians. They returned to the original position, which is about 0.75 m (2.5 ft.) from the upstage wall, in three rows for a total of 18–25 singers. This was the preferred location because it improved choral cohesion and supported the basses.

4. All the banners were gradually stored with increasing liveness and resonance as each banner was stored. All the banners stored had the best sound. It was still clear and bright with some supporting resonance.

Piano Soloist—Colleen Lee

1. The upstage or downstage location for the piano was explored in detail. Close to the upstage, the wall was warm but made the high frequencies brittle. Near the downstage, the stage edge was deemed too harsh sounding. A mid location on center line, 2.8 m (9.1 ft.) from the edge of stage, was deemed ideal.

2. The ceiling #1 located farthest upstage was made steeper, from 9° to 13°, because there was a perceived difference in presence in the seating area behind row P. This was clear when the hall's wall banners were deployed and less so when stored. The angle was increased for this ceiling, and the presence increased as our calculations predicted. Ceiling #1 remained at 8.1 m (26.6 ft.) off the stage.

3. All the stage banners were stored including the upstage drape. The house banners were adjusted in various locations. The preferred position, at least for rehearsals, was to have the rear three sidewall banners deployed and each side and the two closest to the stage stored.

Amplified Guitars, Drums, Bass, and Singers

1. All the drapes and banners were fully deployed. The hall was free of any echoes or resonances with the rented-in speaker stacks played at between 90 and 95 dB. The hall was very well controlled, and the speech clarity was excellent. Even loud segments did not cause the hall to overload excessively or get "harsh."

2. Note that, at the house mix location, the mix console was set at the lower-level step, and the operator was on the next step. This caused him to work on a small crate lid, which was quite uncomfortable. The console should be raised so that the operator can be seated at a normal level.

3. I was told that a number of the musicians in the show were interested in returning to record in the space as they liked the sound very much.

Chamber Music Concert with Video and Narration (City Chamber Orchestra of Hong Kong [CCOHK]) Rehearsal, December 6–7, 2012

1. The orchestra was repositioned before the start of the rehearsal to bring the percussion to the left side of the stage because all the musicians were crowded in the center of the hall. Balances between instruments were generally acceptable, though the brass musicians were complaining that they could not hear the strings well.

2. The orchestra was tightened up. The strings were moved upstage 1–1.5 m (3.3–4.9 ft.), and the brass was moved downstage. This improved hearing between sections. The brass would have preferred a modest riser system, but this may be more psychological and physical. To my listening, the cross stage hearing was very good.

3. Clarity and presence were excellent with the ceiling reflectors providing onstage hearing and overhead presence energy for the first 8–10 rows. When the stage drapes were stored, the stage sidewalls filled in missing reflections to the back half of the room. Stage drapes, when stored, added color and depth to the sound. House drapes, when stored, added modest liveness.

4. The hall was never overly reverberant or highly immersive, even with all the drapes stored. The modest reverberation and steep seating rise create a somewhat frontal sound, and there are crispness and clarity that are very pleasing. The individual instruments were clearly audible. The second clarinet was easily separated from the first—a rare feature in a hall. It was loud at some point but never overbearing or harsh in timbre or color. The strings sounded cohesive and well blended, and the cello and bass had a warm, rich sonority. The brass was loud but tonally even and distinct with the low brass (tuba) never getting muddy or boomy.

London Festival Opera Rehearsal and Performance, December 5, 2012

1. We started upstage with the piano close to the upstage wall, but it sounded muddy and flat. All the drapes were deployed, which deadened the stage too much, but listing tests proved that the ceilings were sending overhead sounds clearly. There was no need to adjust.

2. The piano downstage, just on the downstage side of the stage extension on short stick, sounded much better. Vocalists moved downstage about 3 m (9.8 ft.) from the stage edge on center line.

3. It was very responsive, bright, and clear. According to Christopher Gould, the pianist, it was a "lovely sound." They are very happy with the response from the hall and the ability to sing very quietly. They liked the feedback that they got from the hall, even down in the front rows.

4. In rehearsal mode, the last two drapes on either side of the house should be deployed left. They should be withdrawn during the performance.

5. The piano position for soloists and the accompaniment was recorded at 3.475 m and 5.165 m (11.4 ft. and 16.9 ft.) from the stage edge.

6. The large stage with full extension has the fullest and richest sound for soloists and should be the default and "normal" condition for all acoustic productions, except when there is a strong need for additional seating.

Appendix VII: Michael and Susan Dell Hall, Long Center for the Performing Arts, Austin, TX: A Hall Created within the Bones of a 1950s Auditorium

Building Blocks

The Long Center for the Performing Arts in Austin, Texas, is located in a community that enthusiastically embraces traditional arts and cutting-edge performances. The Austin Lyric Opera needed a new home in which to flourish. Local experimental theater groups such as the Rude Mechanicals, known for performances with comedy and nudity, needed a home too.

These groups and many community members demanded technologically sophisticated spaces that would function exceptionally for both natural sound and highly amplified events. The city of Austin donated an acoustically challenged civic auditorium to a new nonprofit corporation chartered to build and operate a new arts center.

The partnership between the city, the Long Center's board, and numerous community and civic groups faced many challenges. It took many years and a false start to become the privately funded Long Center (Figure A7.1). In a city known by longtime residents to be terminally democratic, it is remarkable that the hall opened at all, let alone on time and on budget. The stakeholders made community engagement a top priority, and as a result, Harley-Davidson leather jackets mingled with ball gowns and tuxedos with startling ease on the opening night.

The Long Center is a community performing arts center that is the principal home of the Austin Symphony, Austin Lyric Opera, and Ballet Austin. It is also the home to touring road shows, Broadway productions, and community events. The center replaced and incorporated much of Palmer Auditorium, built in 1959, and overlooks Lady Bird Lake.

Figure A7.1 Dell Hall at Long CPA, Austin, TX, 2008. A 2400-seat multi-use hall with highly flexible acoustics.

Challenge

The challenge for this facility was to design an acoustically excellent hall with a large capacity of 2442 seats, with only two balconies and within as much of the original auditorium as possible. Our team explored every possible technique to stretch the budget in order to achieve a true multi-use hall.

Solutions

Our innovative solutions included the incorporation of sound-transparent balconies on both levels, placement of acoustic volumes behind throat walls to lower the required roof height, and stepping of side walls to enhance early reflections. The rear of the hall expands in width to bring audiences closer without sacrificing acoustic excellence.

Creating the Building

The $77-million project incorporated two venues: the 2442-seat Michael and Susan Dell Hall and the 229-seat Kevin and Deborah Rollins Studio Theater. Most of the original building was either reused or recycled to build the new hall. The stage house was almost completely reused and greatly expanded to meet the needs of the opera company's large set pieces.

An early design with Skidmore, Owings & Merrill, Fisher Dachs Associates (FDA), and Jaffe Holden was deemed too expensive after the economic downturn. Then, a revised and downsized design was completed with FDA, Zeidler Partnership Architects, and Team Haas Architects. The downsized design significantly reduced the public spaces but did not compromise the acoustic design.

The original design had three halls and three balconies within Dell Hall. The revised, cost-conscious design lowered and widened the hall, resulting in two deep balconies to accommodate the 2442 seats requested by the opera, symphony, and touring productions. These wide, deep balconies presented an acoustic challenge, and so our team approached the design by using innovative supplemental side volumes and sound-transparent balconies.

Architectural Details

Width

An interesting acoustic innovation was the extra wide hall width at the rear of the hall in excess of 100 ft. (30.5 m). This was done to compact the audience area and bring the 2442 seats as close to the stage as possible in addition to lowering the ceiling for the same acoustic volume (Figures A7.2 through A7.6). The side wall early C80 reflections are created by carefully shaping the side walls and ceiling in the front third of the hall known as the throat. The walls provide a transition from the proscenium area where the hall is 60 ft. (18.3 m) wide. Gentle steps and curves distribute side wall reflections to the center seating zones evenly. Rear seating areas are rich in lateral reflections from this throat zone and open to the acoustic volume through the sound-transparent balconies for full envelopment. See Chapter 12 for details on throat wall shaping.

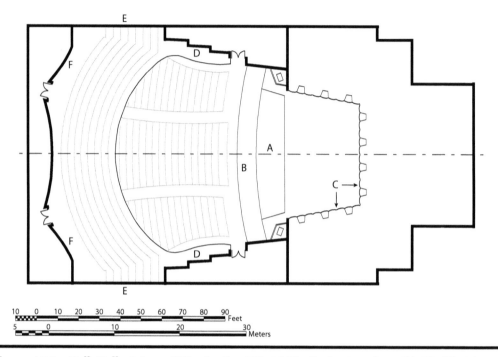

Figure A7.2   Dell Hall at Long CPA, Austin, TX, 2008. Orchestra plan. (A) Pit lift 1/stage extension. (B) Pit lift 2/stage extension. (C) Orchestra shell towers in large shell configuration. (D) Stepped sidewall shaping. (E) Diffusive and parallel sidewalls. (F) Convex-shaped rear walls.

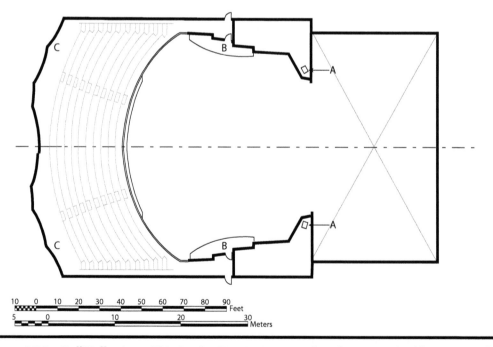

Figure A7.3   Dell Hall at Long CPA, Austin, TX, 2008. Mezzanine plan. (A) Main side array speakers recessed behind grilles. (B) Stepped walls with grilles open to volume behind. (C) Convex-shaped rear walls.

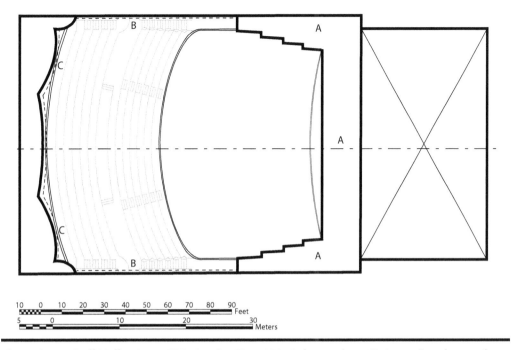

Figure A7.4 **Dell Hall at Long CPA, Austin, TX, 2008. Balcony plan. (A) Acoustic volumes above and around proscenium area. (B) Acoustic drapes vertically drop along sidewalls. (C) Acoustic drapes track horizontally to cover convex rear walls.**

Acoustic Volumes

Additional acoustic volumes were located behind and above proscenium throat walls to reduce the hall's height and costs. Taller walls would mean more blocks, bricks, and steel on the outside, thus increasing cost. These volumes were hidden behind sound-transparent wood grilles above the side box areas near the front of the room. The volume was critical for both symphony and opera, and painted massive block walls served to increase reverberation time (RT). The volume served as a convenient location for acoustic drapes to deaden the sound for amplified events. Deploying drapes in this location meant that they were not visible to the public, thus could be less finished and less expensive.

Adjustable Acoustic Systems

Acoustic drapes made of 24-oz. velour were all motorized and stored in drape pockets when not deployed. They were located primarily in the upper attic volume above the ceiling reflectors with the exception of the cheek and upper rear wall drapes. The range of RT achieved by the drapes and the portable acoustic shell (Wenger Diva type with custom ceiling shape) is quite remarkable.

Balconies

Deep balconies can throttle the sound reaching the back rows of the seats under them. They act as a sound-absorptive chamber that soaks up sound in the hall and reduces the quality of

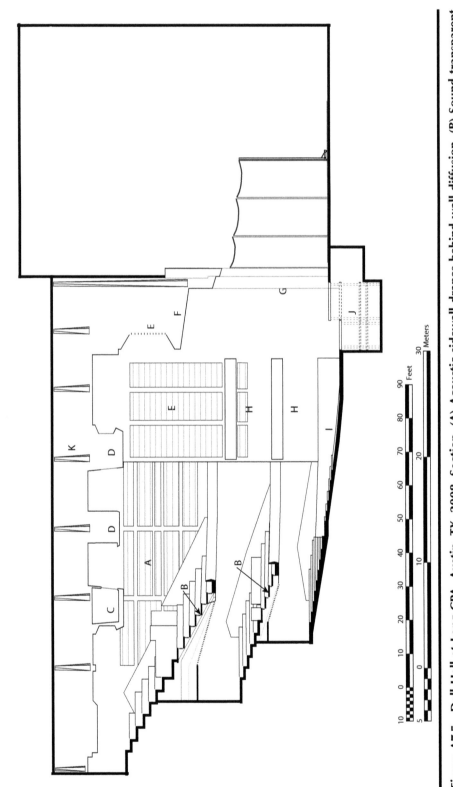

Figure A7.5 Dell Hall at Long CPA, Austin, TX, 2008. Section. (A) Acoustic sidewall drapes behind wall diffusion. (B) Sound transparent balcony both levels (Chapter 11). (C) Spot Booth. (D) Theatrical catwalks. (E) Wall grilles acoustically open to volumes behind. (F) Forestage reflector. (G) Side speakers hidden behind grilles. (H) Box Soffits for lateral C80 reflections. (I) Parterre wall. (J) Double pit lifts/stage extensions. (K) Acoustic drapes in upper hall volume.

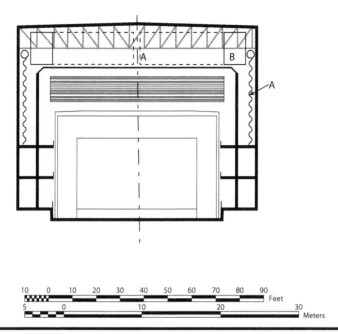

Figure A7.6 **Dell Hall at Long CPA, Austin, TX, 2008. Transverse section. These views are critical to the acoustician. (A) Acoustic drapes behind side volumes and above the ceiling store in pockets (B).**

unamplified sound for the entire audience. This resulted in poor sound not only for a few inexpensive seats in the back of the balconies but also for the entire 2442-member audience. The deep balcony reduced the envelope since less sound returns from the rear of the hall in addition to the G and RT.

The sound-transparent balcony design concept was first successfully used at the Weidner Center for the Performing Arts at the University of Wisconsin-Green Bay in the 1990s to solve a similar issue. During a site visit test, I walked toward the rear of the hall listening carefully to a symphonic performance. I wanted to determine if there was a reduction in surround sound and envelopment as I walked beneath the deep balcony. I repeated the experiment many times—even with eyes closed—and could not detect a difference in sound quality between the open hall and beneath the balcony. Other sophisticated listeners repeated the experiment with the same results at both Weidner Hall and Dell Hall.

The details of the sound-transparent balcony design were simple in concept but difficult to execute (Figure A7.7).

Vertical risers in the balcony were designed for the open position, covered only in metal mesh to keep out debris. The balcony steel was minimized to maintain an open structure. Fireproofed raker beams were boxed in with drywall, resulting in an increase in sound reflection. The ceiling below is a series of vertical wood slats with 60%–70% open area to allow free flow of sound (Figure A7.8). Supply air ducts overhead were boxed in and hug the raker beams to minimize obstruction to sound passing through the assembly. Sprinklers, lighting, conduit, and piping were complicated by this design but well worth the effort. Finally, all areas were painted black in the void space, and most patrons have no idea that it is there. See Chapter 13 for more details on this innovation.

Figure A7.7   Dell Hall at Long CPA, Austin, TX, 2008. In 2007 the structure is clearly visible showing openings for the future sound transparent balcony.

Figure A7.8   Dell Hall at Long CPA, Austin, TX, 2008. Underbalcony view during construction shows ceiling grilles masking the sound transparent openings above it.

Measuring Results

Using the acoustic drapes and with the shell in place, the Long Center's Dell Hall achieves a remarkable swing in RT—from over 2.2 seconds to 1.8 seconds (mid-frequency unoccupied measurements). In amplified performance mode and with the shell removed, the hall drops to 1.6 seconds empty and to 1.4 seconds with an audience present (Figures A7.9 and A7.10).

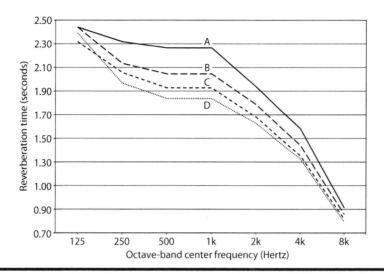

Figure A7.9 Dell Hall at Long CPA, Austin, TX, 2008. RT measurements of unoccupied hall. (A) Drapes and sidewall banners stored. (B) Drapes stored—sidewall banners exposed. (C) Drapes exposed—sidewall banners stored. (D) Drapes and banners deployed.

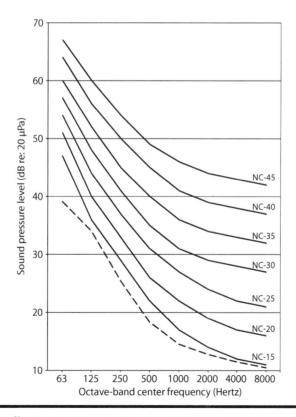

Figure A7.10 Dell Hall at Long CPA, Austin, TX, 2008. NC measurements (dashed line) show levels at NC-15.

Press

Reviews from the local press and the musical community continue to be very positive.

It doesn't take much sound to animate the hall's lively acoustics. A mere two dozen strings in the Dallas Chamber Symphony more than filled the sonic space last week.

Scott Cantrell
Classical music critic, The Dallas Morning News (2013)

Congratulations on designing a space for us where the music and the orchestra can shine!

Peter Bay
Music director and conductor, Austin Symphony (pers. comm.)

Project Information

DESIGN TEAM

Architects: Nelsen Partners
Architects: Zeidler Partnership Architects
Theater consultants: Fisher Dachs Associates

DETAILS

235,000 SF
2425-seat Dell Hall
232-seat Rollins Studio Theatre

BUDGET

$77 million

SCOPE OF WORK

Architectural acoustics
Sound isolation
Mechanical system noise control
Full-featured sound reinforcement system
Adjustable acoustics

CLIENT

Long Center

COMPLETION DATE

2008

Appendix VIII: Nancy Lee and Perry R. Bass Performance Hall, Fort Worth, TX: A Paradigm Shift in Multi-Use Halls

Building Blocks

Bass Hall, located in the Dallas–Fort Worth metroplex, opened to great acclaim in 1998 as the last great hall of the twentieth century. The hall, a unique multipurpose facility, is able to house symphony, ballet, opera, stage, musicals, and rock concerts. It is now the permanent home to the Fort Worth Symphony Orchestra, Texas Ballet Theater, Fort Worth Opera, and the Van Cliburn International Piano Competition (Figure A8.1).

In 2001, the adjacent Maddox-Muse Center officially opened as a rehearsal and recital hall facility, as well as office spaces for various performing arts groups. The center includes the Van Cliburn Recital Hall with seating for 250–300 guests and the McDavid Studio event space with 220 seats. Performing Arts Fort Worth, Inc. is a nonprofit organization that oversees management of the hall and the Fort Worth Symphony Orchestra. Together, they form a thriving downtown performing arts center complex for the fastest-growing city in the United States.

The Will Rogers Memorial Center was the home of the Fort Worth Symphony, the ballet, and the opera since 1936. The 2800-seat art-deco-style Will Rogers Auditorium served as their home for over 50 years, but the acoustics were poor, and volume was low in the wide, fan-shaped room. It is common for halls of this age group to lack early reflections and have undesirable reverberation time (RT) and bass ratio (BR). A large single balcony in the hall also contributed to the poor acoustics.

In the late 1980s, Jaffe Holden and Fisher Dachs Associates conceived a plan for massive renovation of the Will Rogers Memorial Center. However, the regional economy faltered, and the bond issue for the project was not supported by voters.

The Bass family, prominent Fort Worth real estate developers, and other arts patrons stepped forward in the early 1990s with a plan for a privately funded multi-use hall to be the centerpiece of urban revitalization just east of downtown.

Figure A8.1   Bass Performance Hall, Fort Worth, TX, 1998. Bass Hall is reminiscent of a 19th century classical concert hall or opera house, yet features many modern adjustable acoustic systems.

Challenge

A renovation of the Will Rogers faced major challenges. The site was bound by adjacent structures, and the historic art deco lobby and façade compromised the acoustic design. A new hall would not face these restrictions; however, its status as a privately funded enterprise meant that the hall would need to be financially self-sustainable. In order to meet cost restrictions, the new hall would need to have excellent natural acoustics and work equally well for more profitable, amplified events.

Solution

A new hall would need to feature profitable popular entertainment and Broadway productions to offset the reduced rents of the symphony, ballet, and opera. The hall's acoustics needed to be highly flexible and adjustable in order to serve a variety of performances. The classical aesthetics of the hall demand a new paradigm in adjustable acoustics because there would be no obvious location for adjustable acoustic treatments such as banners and drapes.

The acoustic design offered opportunities and challenges that Jaffe Holden approached in new and innovative ways. Bass Hall is one of the most flexible halls on which I have had the pleasure to work. It is still impressive that the hall can transform from an orchestral configuration to a bare stage in a matter of minutes—not hours. Recently, Bass Hall was voted by *Travel + Leisure Magazine* as one of the top 10 best opera houses in the world (Daniel 1999) (Figures A8.2 through A8.5).

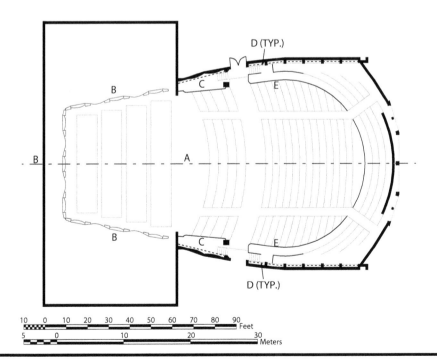

Figure A8.2   Bass Performance Hall, Fort Worth, TX, 1998. Orchestra level plan. (A) Orchestra pit lift and stage extension. (B) Orchestra shell towers. (C) Throat walls with convex shaping. (D) Acoustic banners pull up over sidewalls. (E) Parterre rail.

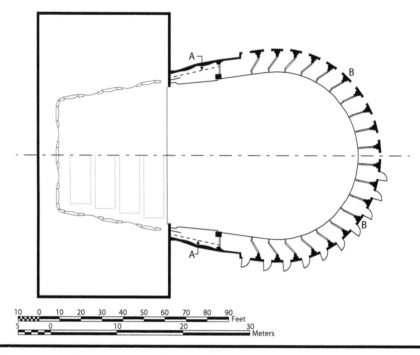

Figure A8.3 Bass Performance Hall, Fort Worth, TX, 1998. Box tier plan. (A) Acoustic drapes cover throat walls. (B) Box tier walls are diffused with applied moldings and columns.

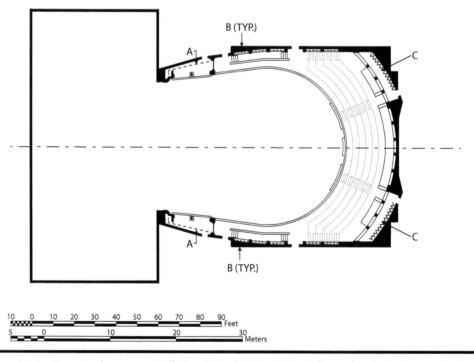

Figure A8.4 Bass Performance Hall, Fort Worth, TX, 1998. Balcony level. (A) Acoustic drapes on sidewalls. (B) Acoustic banners pull up from wood enclosures. (C) Acoustic drapes on rear walls store in pockets with operable doors.

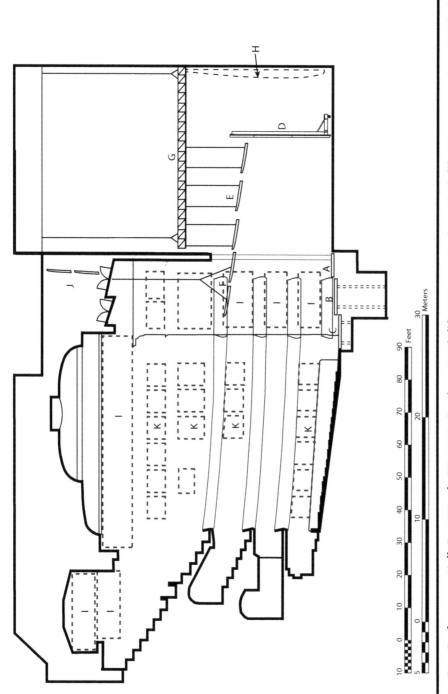

Figure A8.5 Bass Performance Hall, Fort Worth, TX, 1998. Section. (A) Pit lift 1/stage extension. (B) Pit lift 2/stage extension. (C) Platforms cover extra-large pit opening. (D) Shell towers with tuning doors. (E) Acoustic reflectors fly in from upper ceiling. (F) Forestage reflectors store above hall ceiling (J). (G) Upper ceiling of Concert Hall Shaper closes off stage house volume. (H) Acoustic tuning drape on upstage wall. (I) Acoustic horizontally tracking drapes. (K) Acoustic banners pull up on brass rail guides from motors in attic.

Building the Building

The team of Jaffe Holden, Fisher Dachs Associates, and David M. Schwarz (DMS) collaborated on a world-class hall for the owner, Edward Bass, and his family. Ed was involved in every step of the design aided with project management by Theatre Projects Group. It opened in 1998 and was an immediate success. Seventeen years later, it is still considered a great hall.

Architectural Details

Concert Hall Shaper

An ingenious concert hall shaper orchestra shell was implemented in the new hall. This shell was first implemented at the Tokyo International Forum. As described in Chapter 9, the shell was designed to enhance symphonic acoustics on stage and aid rapid stage changeover.

Ceiling and Reflectors

The hall featured a closed dome-shaped ceiling and necessitated the design of new adjustable acoustic wall banner, drape systems, and a large deployable forestage reflector that could store above the closed dome ceiling.

The closed ceiling design by architect David Schwarz was unique because it stipulated a tall forestage area that lacked permanent low-level sound reflectors and catwalk positions. The tall ceiling height was correctly angled and shaped for ballet and opera but was too tall for symphonic acoustics. We recommended two swoop-shaped lower acoustic reflectors, or clouds, to support the symphony playing on the apron and lift. The design of this large moving element required a very high level of coordination with the team. The tunable gold-leafed wood reflector is rigged to drop from a storage area above the ceiling. It descends on counterweighted rigging through a series of operable ceiling hatches that close to form a solid ceiling when the reflector is in position.

The need for stage lighting complicated this design. The play position is about 30–35 ft. (9.1–10.7 m) above the stage and provides onstage reflections to musicians as well as early overhead reflections to the orchestra seating. These fall within the 20–30-milliseconds ITDG and contribute C80 reflections to the center orchestra seating zone.

Shaper Ceiling

The unique shell ceiling and the acoustic volume surrounding the inner shell are crucial components of the outstanding symphonic acoustics of Bass Hall. The lower ceiling provides sound projection and critical early reflections that aid onstage hearing. The contained upper volume provides reverberant energy (late RT, enhanced BR) to musicians and the audience closest to the stage. The whole multiton assembly stores on the upstage wall of the stage house, clearing the rigging of shell ceilings obstructions. A large apron and a lift bring most of the orchestra's string sections in front of the proscenium arch and allow full communication with the hall's large acoustic volume plus C80 and D50 reflections.

Adjustable Acoustic Systems

Bass Hall incorporates many innovative adjustable acoustic systems that are described in detail within Chapter 15 plus others rarely used before. For example, the orchestra shell towers include dozens of operable tuning doors that allow measured amounts of orchestral sound energy to move in and out of the acoustic volume behind the towers. Doors are preset in configurations based on both instrumented measurements and our listening experiments to control the amount of sound that enters the surrounding volume. This ensures that the most pleasing sound reaches both musicians and the audience. An acoustic tuning drape is located on the upstage wall of the outer volume to control the overall reverberance of the volume behind the towers.

The architects did not want to mar the finished plaster ceiling nor tamper with the hall's nineteenth-century feel by using sound-transparent grilles or mesh. This meant that there was not a location for acoustic attic drapes—all banners and drapes would need to be located below the shaped ceiling of 1 in. thick plaster. To overcome this challenge, decorative side wall banners and drapes were located upon every available wall to provide more absorption and to reduce the hall's RT from over 2 seconds (mid-frequency occupied) in symphony mode to 1.5 seconds for amplified productions. Fully lined velour drapes on motorized rails store in pockets or attic volumes behind motor-operated doors. In locations where entrance doors would interrupt the track of drapes on lower walls, double-thick velour acoustic banners rise from handsome wood credenzas on brass rail guides (Figure A8.6).

Figure A8.6   Bass Performance Hall, Fort Worth, TX, 1998. Adjustable acoustic banners of two layers of velour rise from wood enclosures on brass guide rails. Motors in attic operate the drape systems with aircraft cables.

Holistic Approach to Shaping

Contemporary architects often prefer a clean interior design and do not want to include adornment that would be intrinsically acoustically diffusive. David Schwarz's classical design detailing and ornamentation worked in concert with our acoustic recommendations for this project. They promoted the diffusion shaping we wanted, reminiscent of a classical hall like Carnegie Hall or a nineteenth-century European opera house. I referenced Wiener Musikverein and Boston Symphony Hall's columns, panels, and sculptures (putti) as examples of ideal diffusive elements. For the first time in my career, I was not laughed out of the room.

Shaping

On the side walls, round and square wood columns and panelized wall elements provided the ideal diffusion in the scale discussed in Chapter 12. One compromise by Schwarz was to include gentle convex curves on critical throat walls. These spread out early D50 throat wall reflections that might otherwise have been trapped behind decorative columns. Convex curves were not part of their vocabulary. The subtle shaping is nearly invisible to the eye but makes a significant impression on the ear.

Soffits and Balconies

Multiple soffits and shaped balcony fronts of the opera house–styled room realized ideal opportunities for reflection and diffusion surfaces. While plaster was used for many modeled surfaces including balcony fronts and underbalcony soffits, we suggested that the construction be composed of one, two, and three layers of drywall adhered and nailed to concrete block. This recommendation served as a cost-saving measure and insured a BR of 1.3–1.4. The layering created the architects' desired paneled design effect and insured high-frequency diffusion (Figure A8.7). This is another example of how the architecture and the acoustics merged in a holistic manner.

Dome Ceiling

When DMS first suggested a domed ceiling, I was surprised. Extreme focusing, hot spots, and uneven sound distribution often occur in domed spaces. Halls well known for disastrous acoustics include the Royal Albert Hall in London and Duke University's Baldwin Auditorium (prerenovation) in North Carolina. However, the owner and the team were adamant that we make the dome work.

I asked myself how to make lemonade out of this lemon. How would we alter the focal point, diffuse the surface, add convex elements, and perhaps use the dome to actually enhance diffusion?

In order to remove the curse of a focal point, I suggested the use of the French curve with constantly varying radii. This is called a parametric curve or, more accurately, the quadratic Bezier curve B(t). This is defined as

$$B(t) = (1-t)B_{P_0, P_1, P_2}(t) + tB_{P_1, P_2, P_3}(t),\ t \in [0,1].$$

where P_0, P_1, P_2, and P_3 are points outside the curve, and t ranges from 0 to 1.

Figure A8.7   Bass Performance Hall, Fort Worth, TX, 1998. Concave box tier walls are shaped for diffusion using applied moldings, columns, and setbacks in the drywall laminated to the concrete block. An excellent example of a holistic acoustic design by David Schwarz Architects.

The new curve was much less focusing but still created issues with energy concentration. We broke the large curve into two quadratic Bezier curve–shaped domes, which the architect liked. Then, we added radial diffusion elements, or ribs, to the dome. The architect did not really like these. Finally, a light fixture element was needed for room illumination. We suggested a convex-shaped glass acoustic diffusor at the center fit with the lighting designer's concept. The visually stunning dome was holistically designed to be acoustically enhancing and to achieve a number of technical lighting requirements (Figure A8.8).

Measuring Results

Bass Hall looks and feels remarkably intimate for having almost 2000 seats. When people first enter the hall, it feels like a 1200-seat room. This visual intimacy is mirrored acoustically. The hall is remarkably intimate, alive, and intense for natural acoustic performances. This is a result of adhering to criteria detailed in this book regarding volume, shaping, materials, orchestra shells, and adjustable acoustic systems.

Throat walls flare from 70 to 80 ft. (21–24 m) in gentle curves that deliver early reflections from musicians both in the shell and out on the single orchestra lift within 30–50 milliseconds to much of the audience area. Underbalcony areas are limited to four-to-five seating rows on the main floor (depending on location) and seven in the mezzanine. Diffusion materials working across the frequency range are well distributed on walls and ceilings. Adjustable drapes and

Figure A8.8   Bass Performance Hall, Fort Worth, TX, 1998. The iconic domed ceiling, a potential sound focusing element, was reshaped to become an acoustic asset.

banners successfully control the room's RT in the mid and high frequencies. The unique orchestra shell is a success acoustically, visually, and operationally. Background sound levels from mechanical systems are inaudible, and sound isolation to the exterior is excellent.

Bass Hall has achieved remarkable success over the last 17 years for all types of performances. Our objective acoustic measurements support the subjective responses of audiences and performers (Figure A8.9).

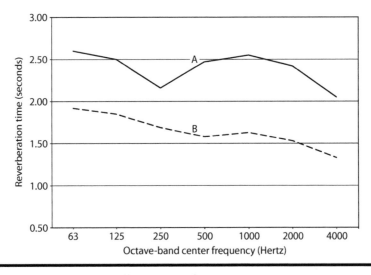

Figure A8.9   Bass Performance Hall, Fort Worth, TX, 1998. RT measurements show an outstanding range of adjustability. (A) Unoccupied, drapes 7/8 deployed. (B) Occupied, drapes fully deployed.

Press

The performance showed off the orchestra's tremendous potential in its new home, with the expanded string section sounding lustrous in a perfect acoustical environment.

Wayne Lee Gay
Nancy Lee and Perry R. Bass Performance Hall (n.d.)

Bass Performance Hall is one of those rare halls in which the music heard by the audience is the same as that heard by the performer. The clarity of sound heard throughout the entire range, in addition to the warm, welcoming environment, makes Bass Performance Hall one of the very best.

Yo-Yo Ma
Nancy Lee and Perry R. Bass Performance Hall (n.d.)

Bass Performance Hall dazzles audiences with its artistry and acoustics.

Southern Living Magazine
Nancy Lee and Perry R. Bass Performance Hall (n.d.)

I've always been impressed with the great concert halls of Europe that are hundreds of years old; but lo and behold, Bass Performance Hall in Fort Worth has surpassed every concert hall in Europe. What a great success!

Tony Bennett
Nancy Lee and Perry R. Bass Performance Hall (n.d.)

A superb piece of urban design and a wonderful place to hear a symphony or see a play.

The Washington Post
Nancy Lee and Perry R. Bass Performance Hall (n.d.)

Breathtaking. It's one of the most beautiful halls I've ever seen.

Judy Collins
Nancy Lee and Perry R. Bass Performance Hall (n.d.)

It has already met or exceeded our wildest expectations regarding the quality of its sound and the beauty of its architectural aesthetics.

Fort Worth, Texas Magazine
Nancy Lee and Perry R. Bass Performance Hall (n.d.)

Bass Performance Hall is the perfect coupling of a beautiful aesthetic with wonderful acoustics.

Renée Fleming
Nancy Lee and Perry R. Bass Performance Hall (n.d.)

Project Information

DESIGN TEAM

Architect: David M. Schwarz Architects
Architect of record: HKS
Theater consultant: Fisher Dachs Associates

DETAILS

2056-seat multi-use hall

BUDGET

$77 million

SCOPE OF WORK

Architectural acoustics
Sound isolation
Mechanical system noise control
Adjustable acoustics
Concert enclosure consultation
Sound reinforcement system

CLIENT

Performing Arts Fort Worth

COMPLETION DATE

1998

Appendix IX: Richmond CenterStage, Richmond, VA: Historic Movie Palace's Acoustic Transformation

Building Blocks

The Richmond CenterStage originally opened in 1928 as a grand movie palace named Loew's Atmospheric Theatre. The theater featured vaudeville shows as well as screenings of the latest Hollywood films. It was designed by the prominent New York architect, John Eberson, who was famous for inventing the atmospheric theater design in which the theater walls resembled an elegant villa or streetscape under a night sky. The design evoked a Spanish setting with a faux sky ceiling containing stars and moving clouds. Modestly renovated in the 1980s, the hall's acoustics for symphony and opera were less than desired. In 2004, Jaffe Holden began plans for the acoustic renovation of the historic Loew's Theatre (Figure A9.1). The initial steps included taking acoustic measurements and attending Richmond Symphony performances.

The theater was renovated in 1983 as a 2200-seat multi-use performing arts center known as the Carpenter Center with an enlarged stage house and ineffectual orchestra shall. The stage rigging system included a noisy 1980s hydraulic pump and motor system that dripped hot oil on stage and frequently failed to operate. The stage lift would drift lower during performances as oil leaked from the hydraulic cylinders. The poor symphonic acoustics were not addressed in the renovation. The tired venue was closed in 2004.

Challenge

The challenge was to transform the once-majestic movie house into a modern multi-use hall while respecting and conserving the glorious interior (Figures A9.2 through A9.4).

Figure A9.1 Richmond CenterStage, Richmond, VA, renovated 2009. Historic character of the magnificent atmospheric design was maintained during the extensive renovation. The orchestra shell is being set and stage extension raised in this photo taken at the tuning. Acoustic clouds are in symphony position. Electronic enhancement system is all but invisible.

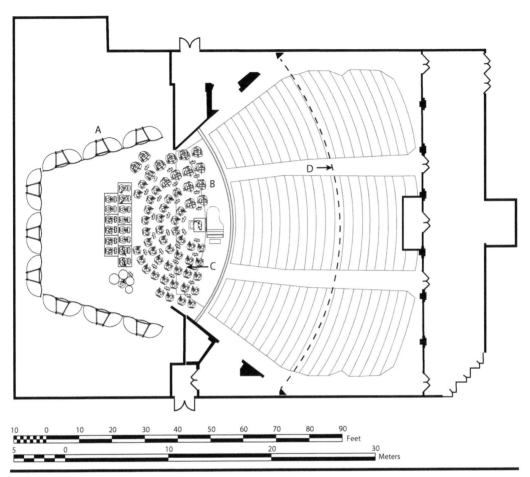

Figure A9.2 Richmond CenterStage, Richmond, VA, renovated 2009. Orchestra plan. (A) New orchestra towers feature sconces and decorative painting. (B and C) Pit lift/extension brings string section out into the hall. (D) Line of balcony front above covers 10 to 11 rows on the orchestra level.

Solutions

Rather than drastically modifying the hall for the Richmond Symphony's acoustic requirements, an electronic solution was used that added the desired reverberation and reflections that would have occurred in an ideally shaped hall. The custom design and painted shell complemented the historic motifs. The orchestra was relocated on large lifts far into the hall.

Building the Building

The interior of the historic Carpenter Center featured the signature atmospheric ceiling made of thin plaster. The plaster was a bass absorber and provided little acoustic diffusion. The volume per seat was quite low at 200 ft.^{3} (5.6 m^3) per person, and the decorative concave side walls provided little C80 nor lateral energy reflections. After extensive measurements and listening tests in the Carpenter Center, we determined that there was no practical physical way to raise the

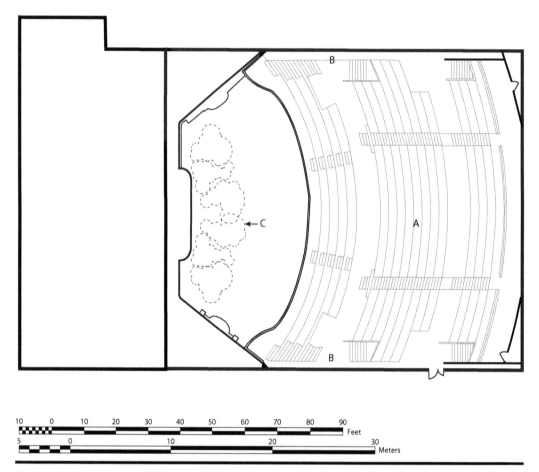

Figure A9.3 Richmond CenterStage, Richmond, VA, renovated 2009. Balcony plan. (A) Balcony overbuild improved site lines and acoustics. Described in detail in Chapter 14, Figure 14.5. (B) Sidewall shaping was diffusive and was left intact. (C) Literal acoustic clouds are flown on winches over the stage extension. They can be raised to the ceiling if required.

mid-frequency occupied RT (1.5 seconds with BR 1.0) or the hall's volume without transforming the plaster ceiling into a sound-transparent metal mesh. Even these construction adaptations would not improve the poor acoustics beneath the balcony. As a result, we broached the subject of incorporating an electronic enhancement system with the symphony and opera. An electronic enhancement system would vastly improve the acoustics without altering the landmarked Eberson architecture.

Architectural Details

Once it was decided that an electronic enhancement system was the best alternative, the design approach forged ahead. The design included using a tunable, flexible orchestra shell plus the most sophisticated enhancement system available. Architectural acoustic improvements were not neglected; sound isolation to the exterior and adjacent spaces was improved, and the rumbling air system was quieted. The pit lift/stage extension was enlarged to bring the symphony further into

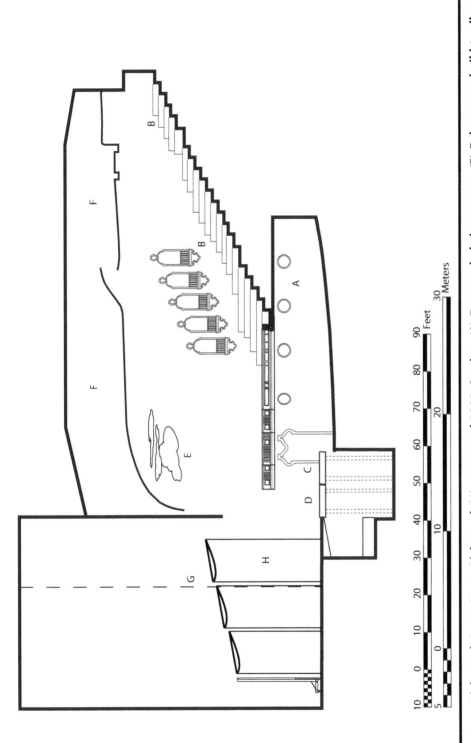

Figure A9.4   Richmond CenterStage, Richmond, VA, renovated 2009. Section. (A) Deep underbalcony zone. (B) Balcony overbuild to allow view of orchestra on lift. (C–E) Acoustic clouds suspended over forestage area (see detail drawing). (F) Volume above historic plaster atmospheric ceiling where EA speakers are located (see Chapter 16). (G) Line of 1928 stage house. (H) Acoustic shell towers and ceilings.

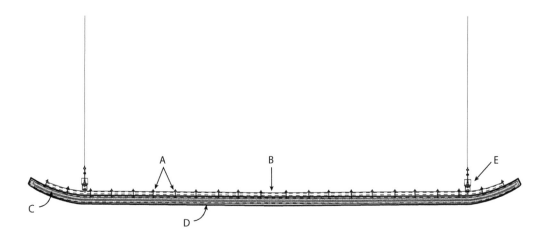

Figure A9.5   Richmond CenterStage, Richmond, VA, 1928, renovated 2009. Acoustic cloud detail. (A and B) Aluminum support system molded into fiberglass. (C) Molded Fiberglass (GFRG) reflector 1 in. (25 mm) thick and bent up at the edges. (D) Surface painted to resemble cloud and angled for improved on-stage hearing. (E) 3 cable supports allow unit to be tuned, descend to the floor for service, or raise to the ceiling to clear productions.

the hall, and the pit was enlarged to support the Virginia Opera. With the orchestra further out in the hall, the late-arriving and focused sound reflections from the concave atmospheric ceiling would have been a hindrance to onstage hearing and orchestral balances. To solve this issue, acoustic clouds, which are literally cloud-shaped reflectors to match the atmospheric motif, were winch-mounted in the ceiling zone over the stage extension. They drop down to an appropriate height of 35 ft. (11 m) and follow a concave arc for symphonic performances to furnish average reflections for blending onstage hearing and balances. The clouds retract for opera performances, clearing site lines for theatrical lighting fixtures (Figure A9.5). Theatre Project Consultants designed these and all theatrical systems.

Orchestra Shell

The architects desired a custom-designed orchestra shell to match the historic Moorish architecture of the Carpenter Center. We had a modest budget and a small area for storage. The result was to modify the standard Wenger Diva towers with a custom paint finish and applied light fixtures.

The historic proscenium's upper corners had decorative architectural elements that blocked the ability of sound to freely move from the stage to the hall. Additionally, the 24 ft. (7 m) high proscenium arch was low and narrow. It throttled sound from moving easily from the stage to the hall. By moving the orchestra forward, the string sections were in front of the arch, and the situation was improved. Deepening the stage extension into the hall was not practical because balcony site lines would be compromised. Electronic enhancement systems significantly improved acoustics for the stage and the hall.

Reraking the Balcony

The reconstruction of the balcony was another acoustic challenge. We needed to improve the sightlines of the extended stage and add much-needed leg room. Seating would be reduced from over 2200 to 1760 so that comfort, sightlines, and acoustics could be improved. Fewer people in the audience resulted in lower absorption and slightly higher RT.

Balconies in Vaudeville houses of this era were designed for audiences that were substantially shorter and of a slimmer physique than modern audiences. This is a common challenge of historic restorations. Adding a new concrete floor is seldom structurally acceptable, as the weight of the slabs would compromise the balcony. Demolishing the balcony was ruled too costly and disruptive. The only option left was to build on top of the existing balcony floor slabs. With concrete out of the question, we needed a lightweight, fireproof material that would be acceptable both structurally and acoustically. Massive structures help ensure a solid BR and a rich, warm RT for a symphony performance. The client was concerned that the floor would sound hollow and feel cheap if it was too lightweight.

The solution was a clever composite of steel studs, decking, and damped plywood. The noncombustible steel decking on steel studs satisfied fire codes, and the fire-treated multilayer plywood decking achieved sufficient acoustic mass. The interstitial damping layer between the plywood sheets used in wood boat constriction to limit engine noise and vibration. This layer deadened footfall noise and made a solid-sounding and -feeling floor. Linoleum floor finish under the seats decreased maintenance, and carpet was limited to the main cross aisles and steps (Figure 14.5).

Electronic Enhancement System

Jaffe Holden pioneered the design of electronic systems to increase RT, C80, lateral reflections (D50), and BR starting in 1967 with the first patented Electronic Reflected Energy System (ERES). Additional patents were added in 1980 for the Reverberation on Demand System (RODS) (see Figure 16.2), which added stability and lowered coloration of the ERES. Dozens of halls in the United States and Japan were outfitted with ERES systems over the last 40 years primarily in renovations. After an extensive study of the program and the architecture, we selected a system for the Carpenter that was a hybrid of the ERES and LARES reverberation system developed by Griesinger and Barbar. For more details, refer to Chapter 16.

Measuring Results

In 2007, the leadership accumulated the $63 million needed to begin the renovation of both the Carpenter and the former Thalhimers department store (circa 1939) adjacent to the theater. In 2009, a new 179,000-ft.^{2} complex became home to four venues:

1. *Gottwald Playhouse*: a 200-seat flexible black box theater with full box-in-box acoustic isolation

2. *Genworth BrightLights Education Center*: a training space with classrooms and offices for young actors, artists, and musicians
3. *Rhythm Hall*: a multipurpose venue used as a gallery and a party and reception space heavily booked for weddings
4. *Carpenter Theatre*: featuring an enlarged stage and forestage, new shell, cloud reflectors, and the most advanced electronic enhancement system available

The Carpenter Theatre's renovation is a complete acoustic success for a full range of amplified music acts and Broadway shows. It serves as the touring home of the Virginia Opera (based in Norfolk, Virginia) and as a home to the Richmond Symphony. Broadway and amplified shows are flattered by the dry (RT 1.5 seconds) existing hall without the use of the enhancement system. For the four opera performances a year presented in the Carpenter, the enhancement system is set to a low-setting RT of 1.7 seconds, and early reflections are added for increased intimacy and clarity. The electronic system is set to either 1.9 or 2.1 seconds of reverberation for Richmond Symphony performances depending on the repertoire and the music director's preference. The underbalcony acoustics are vastly improved, and the early reflection system adds strong lateral reflections between 30 and 80 milliseconds that would not otherwise occur from the wide fan-shaped side walls. A further enhancement is that the orchestra shell has electronic enhancement to improve the musicians' sense of richness and bass resonance that would not be present otherwise.

The RT curves show the unoccupied reverberation time in the hall's natural acoustic condition to be 1.5 seconds at mid frequency, increasing to a maximum of 2.1 seconds for the romantic symphonic repertoire (Figure A9.6). These three presets of 1.7, 1.9, and 2.1 seconds are set with low-frequency RT increasing to 1.20 times midband RT below 180 Hz (BR of 1.2) and are completely natural sounding. The enhancement system is an important component of the hall's success.

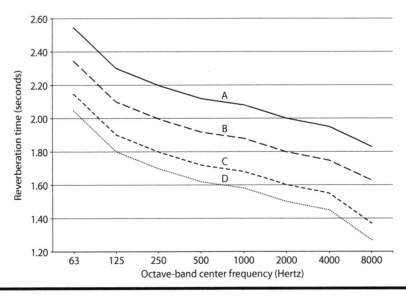

Figure A9.6 Richmond CenterStage, Richmond, VA, 1928, renovated 2009. Electronic Architecture system allows wide range in RT settings. (A) Orchestra setting, long. (B) Orchestra setting, medium. (C) Orchestra setting short. (D) Enhancement system off.

Press

Thanks, in large part, goes to the collaboration between the experienced professionals at Jaffe Holden and our partners at Richmond CenterStage. We have been thrilled to work with Jaffe Holden …

David J. L. Fisk
WTVR (2009)

Orchestral sound in this hall is significantly more reverberant—a short, loud chord takes nearly three seconds to decay to silence, about one second longer than before the renovation. High-frequency instruments, such as flute, oboe, trumpet, violin and cymbals, sound noticeably brighter.

Clark Bustard
Virginia Classical Music Blog (2009)

In Berlioz's Le Corsaire Overture, Smith adopted a tempo that was animated but still broad enough to allow stuttering woodwind figurations to emerge with complete clarity. The string sonorities that he obtained in the Berlioz, and subsequently in Beethoven's Emperor Concerto, had warmth and heft without turning mushy …

Clark Bustard
Virginia Classical Music Blog (2010)

Project Information

DESIGN TEAM

Architects: Wilson Butler Architects
Theater consultants: Theatre Projects Inc.

DETAILS

1760-seat multipurpose theater (Carpenter Theatre)
250-seat flexible theater (Gottwald Playhouse)
150-seat open space arrangement (Rhythm Hall)
Rehearsal spaces
Digital art learning center
Art gallery

BUDGET

$63 million

SCOPE OF WORK

Architectural acoustics
Sound isolation
Orchestra shell design
Mechanical system noise control

Electronic architecture systems design
A/V systems design

CLIENT
Virginia Performing Arts Foundation

COMPLETION DATE
2009

Appendix X: Wagner Noël Performing Arts Center, Midland, TX: The Star of West Texas

Building Blocks

The Wagner Noël Performing Arts Center is located on the campus of the University of Texas of the Permian Basin (UTPB) between the cities of Midland and Odessa. The 109,000-ft.^{2} (10,126 m^2) space was designed to accommodate touring Broadway shows, touring popular music, the Midland-Odessa Symphony and Chorus, dance, and a broad range of music programming. In addition to the 1850-seat multi-use hall, the building includes a flexible recital and banquet hall that seats 200, as well as new music rehearsal rooms, practice rooms, and offices for the UTPB music department (Figure A10.1). The project opened in 2011 and cost $81 million funded primarily by the Texas Legislature and local fundraising efforts.

Challenges

The greatest challenge was to build a highly flexible performance space that would handle a large range of performance types, a flexible recital hall and music education facility on a very tight budget. The local symphony desired an excellent acoustic performance hall but was concerned that they would feel lost in the large hall.

Solutions

The orchestra level seating was configured into two sections, and the symphony extends into the hall on two lifts, as described in Chapter 10. This created an intimate and dynamic acoustic environment on a modest budget returning a significant return on the university's investment.

Figure A10.1   Wagner Noël PAC, Midland, TX, 2009. This 1800-seat hall is shown in symphonic mode. The forestage ceiling is in lowered position, and the orchestra shell is set to full depth with musicians on the lift.

Creating the Building

The Wagner Noël's acoustically flexible auditorium designed by BOORA architects, is terraced with two balconies and stacked side boxes with excellent sightlines from all areas (Figures A10.2 through A10.5). To support the broad range of programming, the hall is equipped with a full fly tower and forestage gridiron, both motorized and manual rigging systems, and an orchestra pit lift with seating wagons. A custom-designed orchestra shell and mover system allows the towers and ceilings to be removed from the stage fly loft when rigging space is required.

For symphony concerts, the orchestra plays out on both pit lifts bringing the orchestra well past the proscenium line, and a forestage piece drops down over the lift area to about 36 ft. (11 m) improving onstage hearing and sound projection (Figure A10.6). The local symphony draws about 1000 patrons, so the upper balcony can be closed and darkened to help make the hall visually more intimate. The enveloping design of the parterre walls on the orchestra level breaks the seating area up to make it feel more populated even when smaller audiences are present.

In order to achieve the target reverberation time (RT), this design required a large acoustic volume of 700,000 ft.^{3} (19,822 m^3) with massive concrete block walls and concrete ceiling surfaces. We specified large vertical double-layer velour acoustic banners as well as traveling horizontal acoustic drapes in the ceiling and on the upper rear walls. The adjustable acoustic drape design was similar to that of the Long Center in Austin, Texas, as described in the case study in Appendix VII, but here we included vertical moving banners that drop between the roof truss steel. This design by Auerbach Pollock Friedlander (APF) was both economical and effective.

The ceiling structure, drape pockets, forestage grid, and return air ducts provide the mid- and high-frequency diffusion, shifting from wall treatments below the ceiling to the suspended

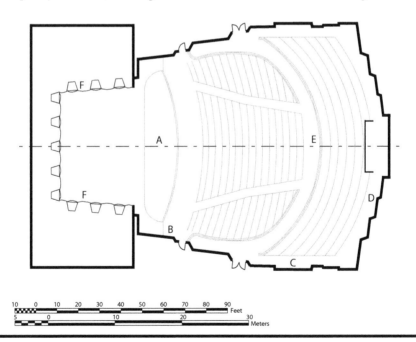

Figure A10.2   Wagner Noël PAC, Midland, TX, 2009. Orchestra plan. (A) Large pit lift/stage extension. (B) Stepped diffused sidewalls at approximately 7°. (C) Sidewall shaping for diffusion. (D) Stepped rear wall prevents echoes to the stage while returning sound energy. (E) Parterre wall. (F) Orchestra shell towers allow modular depth settings.

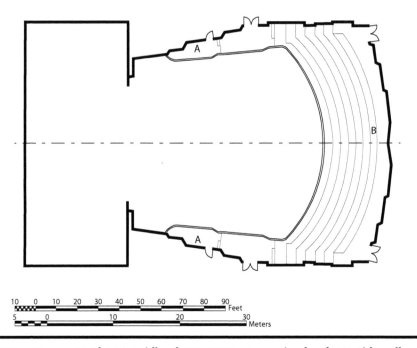

Figure A10.3   Wagner Noël PAC, Midland, TX, 2009. Mezzanine level. (A) Side gallery seating creates soffits for lateral reflections. (B) Seven seating rows under the balcony.

Figure A10.4   Wagner Noël PAC, Midland, TX, 2009. Balcony level. (A) Shaped throat walls extend to sound transparent ceiling. (B) Diffusion shaping with two and three layers of drywall. (C) Stepped rear wall has acoustic drape (shown stored).

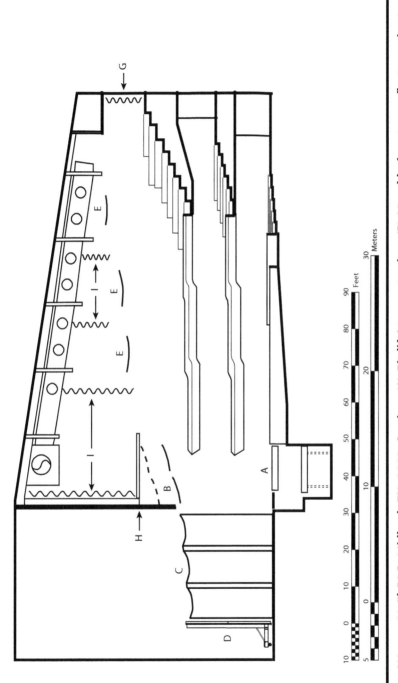

Figure A10.5 Wagner Noël PAC, Midland, TX, 2009. Section. (A) Pit lift/stage extension. (B) Movable forestage reflectors raise to clear lighting angles. (C) Orchestra shell ceilings can be removed to carts to store out of the rigging. (D) Shell towers. (E) Convex-shaped ceiling reflectors at catwalks above sound transparent ceiling. (F) Sidewall acoustic drapes. (G) Acoustic drape on upper rear wall. (H) Vertically deployed double-layer acoustic banners.

Figure A10.6 Wagner Noël PAC, Midland, TX, 2009. The 80-member Midland-Odessa Symphony & Chorale rehearses on stage in the deep shell configuration. The modular shell can be set to varying depths depending on the requirements.

in-space diffusion elements above. Forestage ceiling reflectors are a visual extension of the orchestra shell ceilings and are able to descend and retract.

Architectural Details

Adjustable Acoustic Systems

The vertically deployed acoustical banners are located in the center of the room to drop between the steel truss members that would otherwise block horizontally traveling devices. APF engineered the double-sided 24-oz. wool serge banners, with a 6-in. air gap between the two layers, an efficient full-range absorbing system. Their location reduces the mean free path and works to divide the space into smaller volumes, thereby increasing the acoustical efficiency of the banner. The acoustic drapes on the upper side walls were thick, 24-oz. velour mounted 12 in. (30.4 mm) away from the walls. Adding the air space behind the drapes offers increased low-frequency absorption. By locating them along the upper side walls and proscenium wall, we absorb the lateral energy and reverberation that are developed between large upper parallel surfaces.

Also, the rear acoustic drapes compensate for the lack of an audience in the upper reaches of the hall. Often, we find that to provide the orchestra floor with the correct level of energy results in the upper balcony having a higher level of reverberation in relation to the direct sound and early reflections.

Over 5000 ft.^{2} (454.5 m^2) of double-layer banners were manufactured and installed by Texas Scenic Company of San Antonio. This system has a programmable, preset motor control that was tuned for individual users. Refer to Chapter 17 for a more detailed discussion of tuning this hall.

Orchestra Shell Walls and Ceiling

The design of the orchestra shell walls and ceiling was carefully coordinated with the architect BOORA to complement the audience chamber and achieve the necessary acoustic performance. The modular orchestra enclosure is composed of towers that can be rolled into position. The ceiling reflector panels are shaped in a custom-designed swoosh shape, reminiscent of the Nike logo, to aid onstage hearing and to project sound to the audience. They were also equipped with integrated orchestra performance lighting and then mounted to their own permanent trusses. This allowed them to be removed efficiently as single elements (Figure A10.7).

The orchestra pit lift can be set at stage, audience, orchestra pit, and storage level. At storage level, upholstered seating wagons can be moved to audience level and into place quickly to provide additional seating. At stage level, it can be used as an extension to the stage forming a forestage apron for symphonic performances. At pit level, it forms the orchestra pit for Broadway productions. Adjustable acoustic drapes cover the rear pit wall for tuning and controlling pit loudness and balances.

Figure A10.7 Wagner Noël PAC, Midland, TX, 2009. Ceiling is lowered on a line set to the stage during tuning to verify the correct angles. The angle was adjusted during the hall tuning to increase support of the chorus.

Wall Shaping

The wall elements were designed by BOORA like oversized bricks that modulated in depth and size to provide mid- and high-frequency diffusion but end at the ceiling line. They are formed of multilayer drywall directly adhered to the block walls or on heavy metal studs for RT development and to support the bass ratio (BR). Above the ceiling, which is a totally sound-transparent curtain of glowing LEDs designed to look like stars in the night sky, the masonry walls were painted with three layers of paint to seal the block.

Upper Wall Surfaces

On our first site visit after the upper concrete block walls were painted, the block appeared to be sealed and sound reflective. However, I wanted to be sure that it was correctly treated to be sound reflective. On closer inspection of the flat black paint with a powerful flashlight, I was able to observe the tiny white spots of the block indicating that the block was not thoroughly sealed and therefore would be too absorptive. The solution was to demand another coat of paint, this time applied with a roll, not sprayed. This proved to be difficult and time consuming as the walls in question were large and high up. This second coat provided the complete seal required for the required RT and BR and was confirmed by a round of measurements. We reached a 2.3-seconds mid-frequency RT when the hall was unoccupied.

Measuring Results

The maximum RT mid frequency of the main hall, unoccupied, was measured to be approximately 2.3 seconds, and the minimum RT mid frequency was approximately 1.5 seconds (Figure A10.8). The adjustable absorptive treatments provide 0.8 seconds of variability in the room's response.

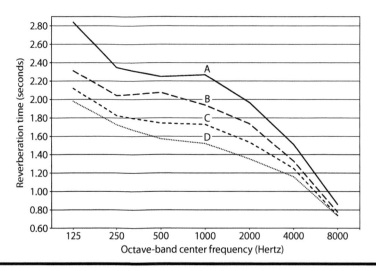

Figure A10.8   Wagner Noël PAC, Midland, TX, 2009 unoccupied. Reverberation time measurement. (A) Banners fully stored. (B) Proscenium and rear banners deployed. (C) Catwalk level banners deployed. (D) Banners fully deployed.

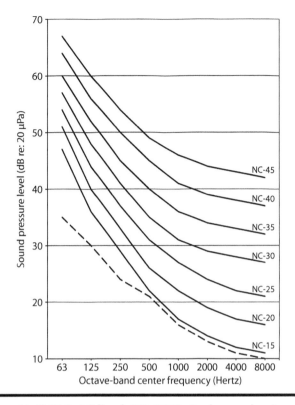

Figure A10.9   Wagner Noël PAC, Midland, TX, 2009. Background noise measurements. Note the underfloor air supply system is inaudible.

When the room is fully occupied, it is anticipated that the RTs will be 0.2–0.4 seconds shorter depending on the number of audience members.

The ambient noise levels were consistent throughout each space. In the main hall, the NC level was 15 at all locations measured (Figures A10.9 and A10.10).

Press

> … As the minutes flowed past the sheer acoustic clarity and loveliness of the Center surrounded us. An instrumental solo can be heard as if you are only a few feet away, while lush layers of sound cascade when the whole orchestra was playing.
>
> **Graham Dixon**
> *In the Limelight—MOSC at Wagner Noël Performing Arts Center (2011)*

Before the Wagner Noël opened, officials hoped the $81 million venue could bring in 200 events a year. Stephanie Rivas, senior marketing manager, said the center now brings in "well over 200 events." In a press conference in February, general manager Ty Sutton mentioned that the venue was 67 in Pollstar's top 100 Theatre Venues for 2013 and the trade magazine Venues Today named it first in 2013 for university venues with 5000 seats or fewer. Rivas attributes Wagner Noël's success to the community she says

Figure A10.10   Wagner Noël PAC, Midland, TX, 2009. Hall tuning. Author listening to choral ensemble in upstage position. This location was not preferred, but it is informative to listen to ensembles in multiple locations.

has embraced the PAC by buying tickets to shows and using the facility for local events and performances.

Midland Reporter-Telegram
Wagner Noël Raises Standard for Arts,
Entertainment in the Basin, published November 3, 2014

Project Information

DESIGN TEAM

Architect: Boora Architects
Architect of record: Rhotenberry Wellen Architects
Theater consultant: Auerbach Pollock Friedlander

DETAILS

Size: 108,500 GSF
1850-seat multi-use hall
200-seat recital hall
Rehearsal room
Teaching suites
Practice rooms

BUDGET

$81 million

SCOPE OF WORK

Architectural acoustics
Mechanical system noise control
Shell design

CLIENT

University of Texas of the Permian Basin

COMPLETION DATE

2012

LEED RATING

Silver

Appendix XI: Wallis Annenberg Center for the Performing Arts, Los Angeles, CA: A New Hollywood Star in Beverly Hills

Building Blocks

Although Los Angeles (LA) is huge and is known as an arts city, it is surprisingly short on multi-use halls. Hollywood and Beverly Hills are well known as the center of the media and motion picture industry, but there are very few facilities where theater, music, and dance can be taught, studied, and performed. LA's despicable traffic situation demands that venues be close to home, and the new Wallis Annenberg Center for the Performing Arts on the site of the historic downtown post office serves this need.

The $70-million, 2.5-acre, site in Beverly Hills, California, is at the intersection of Santa Monica Boulevard and Canon Drive and consists of two buildings—the renovated, expanded 1934 Beverly Hills Post Office that houses the Lovelace Studio Theater and the new 500-seat Bram Goldsmith Theater. The Annenberg Center opened in the fall of 2013 with a gala hosted by Robert Redford, and a host of Hollywood stars were in attendance (Figure A11.1). The multipurpose Goldsmith Theater achieves full acoustic flexibility in a compact form.

Zoltan Pali of SPF:architects, collaborator with Jaffe Holden and Schuler Shook theater consultants, explained his vision to the *Los Angeles Times*' Deborah Vankin: "We were building a new theater next to a historic structure that's not only revered, but a national monument. The proper way to add or put something on this particular site would be to build something completely different and new. In my mind, you have to build for today, and with today's technology and ideas; you also have to make sure that you actually distinguish the new from the old—radically. But you also want to create a dialogue" (Vankin 2013) (Figures A11.2 through A11.4).

Figure A11.1 Annenberg Center for Performing Arts, Los Angeles, CA, 2013. 500-seat multi-use hall features full orchestra shell, pit, sound transparent walls and ceilings, and adjustable acoustic drapes.

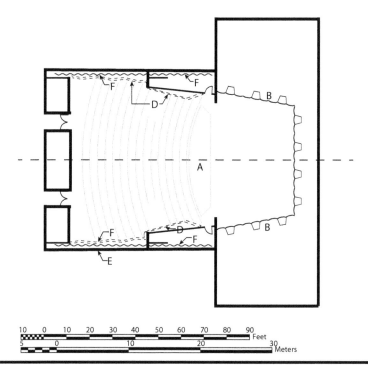

Figure A11.2   Annenberg Center for Performing Arts, Los Angeles, CA, 2013. (A) Orchestra pit lift, motorized. (B) Orchestra shell towers store in upstage zones. (D) Wood slat wall system. (E) PIP thick concrete structural walls behind three-layer drywall walls. (F) Acoustic drapes on sidewalls behind slats store in pockets.

Challenges

Both the owner and the architect wanted the hall to have a wood chamber that was intimate and enveloping. The hall was to feature multi-use programming, so it would require large areas of drapes and banners to provide reverberation time (RT) variability.

Solutions

A wood slat system was engineered to be open enough for sound transparency yet provided a visual sense of solidity and substance desired by the owner. The wood glows from behind with a rich blue LED light. This contributed to the sense of intimacy and envelopment.

Building the Building

Design Process

The 500-seat Goldsmith Theater was conceived as a multipurpose facility to host musical theater, drama, dance, and orchestral and choral performances. Many of these events will use sound reinforcement; however, the room is acoustically designed so that it could be used without sound

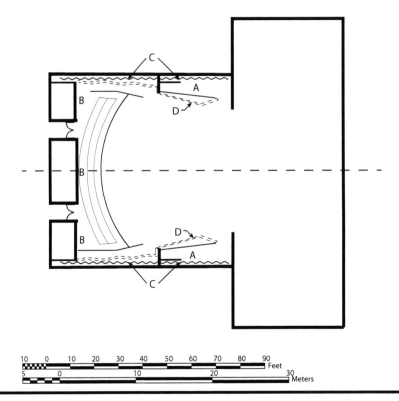

Figure A11.3   Annenberg Center for Performing Arts, Los Angeles, CA, 2013. (A) Throat walls for early reflections. (B) Note shape of rear wall parallel to proscenium eliminates echoes. (C) Adjustable acoustic drapes. (D) Sound transparent wood slat system.

reinforcement, such as for plays without music. The room's natural acoustics are conceived to support unamplified orchestral music such as chamber orchestra and classical soloists.

The Goldsmith was designed to support a variety of programs, and the acoustic systems we proposed to support these programs are described at the end of this case study.

Room Volume and Shape

To support orchestra performances, a volume per seat of 375–425 ft.^{3} (10.6–12.0 m^3) per person was proposed for the auditorium. With a seating capacity of 500 and room for up to 70 musicians on stage, an audience chamber volume of approximately 213,750–242,250 ft.^{3} (6050–6860 m^3) including orchestra shell volume was required to achieve the desired volume/seat ratio.

Dimensions

The width of the theater at the orchestra level is about 70 ft. and ensures advanced acoustical intimacy. Acoustical intimacy is largely determined by the initial time delay gap or the difference between the arrival of the direct sound and that of the first reflected sounds (C80). Additionally, the throat walls at the proscenium were shaped at a +/− 7° slope in the front third of the room. This distributes early reflections to the audience members and accentuates clarity and definition.

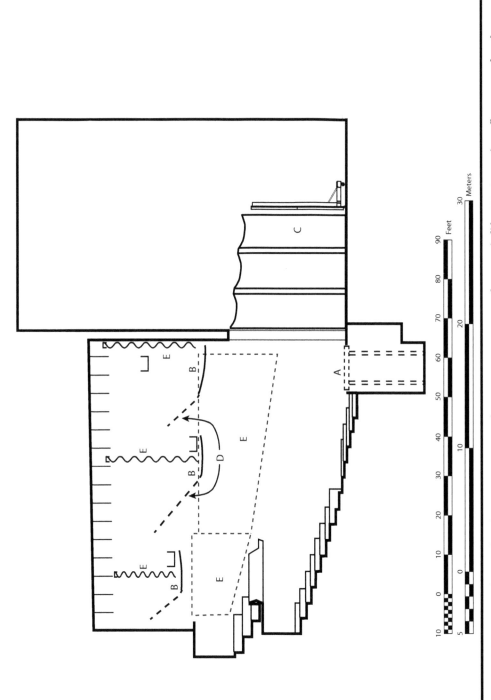

Figure A11.4 Annenberg Center for Performing Arts, Los Angeles, CA, 2013. (A) Orchestra pit lift. (B) Acoustic reflectors under the catwalks. (C) Shell towers. (D) Sound transparent wood slat system. (E) Acoustic drapes on motorized tracks store in pockets.

Figure A11.5   Annenberg Center for Performing Arts, Los Angeles, CA, 2013. Sound transparent wood slats are back-lit with blue LEDs. A cream-colored acoustic drape matches the wall color behind it so the look is consistent.

Adjustable Acoustic Design Ideas

The hall itself hides the adjustable acoustic systems of velour drapes that cover the upper side walls and cross the ceiling on linear tracks. Wood slats line the walls and ceiling and are acoustically transparent and slightly diffusive at high frequencies. These are backed with a fabric scrim that was also tested to be sound transparent. The scrim visually masks the drapes, lighting positions, and other technical systems. Acoustic shaping for diffusion and early C80 reflections also reside behind the wood slats and feature LED accent lighting in swooping curves. White velour drapes over white side walls allow the space behind the wood slats to glow with blue LEDs regardless of whether the acoustic drape is deployed or retracted (Figure A11.5). The LED drivers were carefully tested to be certain that they were quiet.

Sectional Considerations

The steep rise in the orchestra level seating is due to the stipulation that the stage house not be taller than the historic post office. The stage house was buried 30 ft. (9.1 m) below street level. The steep rise is excellent for drama and dance acoustics and offers superb sight lines.

Architectural Details

A custom-modified Wenger orchestra shell completes the stage for classical performances, as described in detail in Chapter 10. We used a dark wood veneer for a rich finish to the towers and ceiling and designed the towers and ceilings to be able to break apart into compact sections that would fit on the stage loading elevator. This was done so that the shell could be totally removed if a large production required more space. The stage shell tower base's store in a very compact way by nesting in recesses carved into the upstage concrete wall.

The Annenberg site is located on a busy intersection near fire and police stations and under the path of frequent private and film studio helicopters. Budget constraints meant that the noisy packaged HVAC systems would need to be mounted on the rooftop over the lobby. In order to achieve the NC-15 criteria, the walls and the roof were formed of thick, 12–18-in. (30–45-cm) reinforced concrete. A 4 in. (10.1 cm) thick floating concrete slab was added to the audience seating area roof to thoroughly block any extraneous sound.

The grand hall of the original post office, now called the Paula Kent Meehan Historic Building, contained many architectural details upon which we worked carefully to stitch them together seamlessly with new aesthetics.

Acoustic plaster was used on the double-curved vaulted ceiling to keep the approximately 6 second RT in the grand hall under 1.5 seconds. Donor plaques hang over what once were post office boxes, and arches that once housed stamp kiosks now contain ticket booths.

Measuring Results

Both the Goldsmith and the Lovelace have been lauded as excellent venues for the programmed performances. The *LA Times* and other local press have praised the acoustics and the flexibility of the facilities (Figure A11.6). A summary of the acoustic measurements taken at the hall tuning is included in Figures A11.7 and A11.8.

Figure A11.6   Annenberg Center for Performing Arts, Los Angeles, CA, 2013. Author working with the University of California Los Angeles (UCLA) Chamber orchestra to determine correct setting for the acoustic systems and ideal stage location based on their feedback.

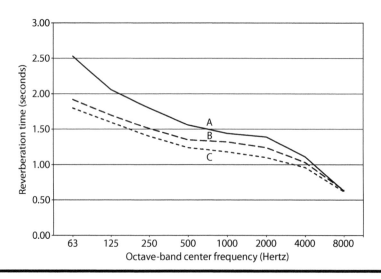

Figure A11.7 **Annenberg Center for Performing Arts, Los Angeles, CA, 2013. Measurements show RT range. (A) Drapes fully stored. (B) Drapes partially deployed. (C) Drapes fully deployed.**

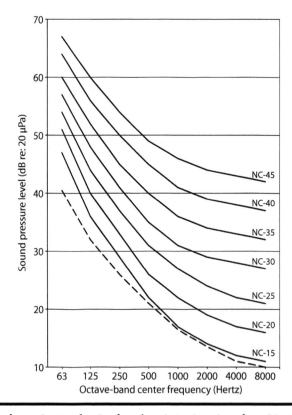

Figure A11.8 **Annenberg Center for Performing Arts, Los Angeles, CA, 2013. NC-15 (dashed line) was achieved in this hall despite the roof-top mechanical equipment and nearby busy city streets.**

Press

The first public performance at the Wallis Annenberg Center for the Performing Arts was held on November 8, 2013, and featured the Martha Graham Dance Company (Swed 2014).

> A hall with many uses—music of all sorts, theater, dance, opera and children's shows—can be an acoustician's riskiest assignment. Amplified music, amplified voice, unamplified music, unamplified voice all require different degrees of reverberation. Music genres and the number of performers make a difference. ... The Wallis' acoustic, designed by the firm Jaffe Holden, is characterized by equally modern, clean, clear and uncolored sonic lines. ... From where I sat, on an aisle mid-hall, the St. Lawrence's crackling electrical energy was evenly distributed from the violins' top notes to the cellos' bottom.

Mark Swed
Music critic, Los Angeles Times (2014)

Project Information

DESIGN TEAM

Architect: SPF:architects
Theater consultant: Schuler Shook

DETAILS

500-seat adjustable acoustics theater
150-seat studio theater
Classrooms
Donor lounge
Lobby

BUDGET

$51 million

SCOPE OF WORK

Architectural acoustics
Mechanical system noise and vibration control
Sound isolation
Concert enclosure

CLIENT

City of Beverly Hills

COMPLETION DATE

2013

LEED RATING

Pursuing LEED Silver

Annenberg Center for Performing Arts Case Study: Table of Acoustic Systems

The following table of acoustic systems was composed to define the criteria for the Annenberg's broad range of programs. It forms the basis of the "Acoustic Manual" used by staff to get the best acoustic results.

Program	Stage Configuration	Pit Configuration	Desired RT	Acoustical Drapes	House Mix	Sound Reinforcement
Chamber orchestra	Orchestra shell	Orchestra plays on pit platform	1.5–1.8	Stored	No	Announce only
Orchestral and chorus	Orchestra shell	Orchestra plays on pit platform	1.5–1.8	Stored	No	Announce only
Choral	Orchestra shell	Chorus plays on pit platform	1.5–1.8	Stored	No	Announce only
Dance	—	Audio playback	1.2–1.6	Partially deployed	No	Announce
Musicals	—	25 musicians max	1.0–1.4	Partially to fully deployed	Maybe	Maybe
Jazz band	May use the orchestra shell	Band plays on pit platform	1.0–1.4	Partially to fully deployed	Maybe	Maybe
Drama	—	Occasional audio playback	1.0–1.4	Fully deployed	Maybe	Maybe
Amplified	—	—	1.0–1.3	Entirely deployed	Yes	Yes

Glossary

Acoustic terms used in this book.

Term	Definition	Equation	Variables
Absorption coefficient α	The amount of sound absorbed by a material indicated by the absorption coefficient α. Ranges from 0.01 (totally reflective) to 1.0 (totally absorptive) and tested in a special chamber designed for this use.	$\alpha = I_a/I_i$	where I_a is the sound intensity absorbed; I_i is the incident sound intensity (W/m^2)
Acoustic baffles	An acoustic baffle is any construction used to mitigate sound. Examples include sound barriers to block sound between instruments or absorptive panels in a room to reduce the sound that is reflected.		
Acoustic banners	In this book, banners refer to vertically moving fabric panels, usually wool, woven mesh, or other sound-absorptive material. They are usually motorized and remotely controlled.		
Acoustic energy	Defined as the amount of sound present (acoustic pressure) in a given volume or space. Higher acoustic energy will be perceived as louder. Boundary surface materials and location will define the acoustic energy.		

Term	Definition	Equation	Variables
Apron	The apron is part of the stage between the proscenium arch or fire curtain and the edge of the stage. Usually, it does not include the orchestra pit lift or stage extension.		
Arbor pit	An arbor pit is an open slot in the stage floor below the rigging wall permitting greater travel for counterweight arbors and pipe battens.		
Articulation	Similar to clarity and immediacy. The ability to clearly hear definition in music such as the attack of the bow on a string.		
Average reflection rate	The average reflection rate is the distance traveled in 1 second (the speed of sound) divided by the average distance between reflections (the mean free path).		
Balance	Balance refers to the relative sound levels of various sections of an ensemble and/or solo artist. Halls should have well-balanced sound with performers that can easily project.		
Bass ratio (BR)	BR helps to quantify the "warmth" or relative strength of the bass response of a space. It is the ratio of the sum of the RT60 of the low frequencies (125 and 250 Hz) to the sum of RT60 at mid frequencies (500 and 1000 Hz).	$BR = \dfrac{RT_{125\,Hz} + RT_{250\,Hz}}{RT_{500\,Hz} + RT_{1000\,Hz}}$	
Binary amplitude diffuser (BAD) panels	BAD TM RPG systems. These panels provide both absorption and diffusion using a two-direction binary amplitude grating behind a sound-transparent fabric.		

Term	Definition	Equation	Variables
Blend	A harmonious mixture of orchestral sounds. Not one instrument stands out, and sound is well mixed and homogeneous.		
Boxes, side boxes	Audience seating areas on the sides or rears of halls, often with 6–8 loose chairs. They create useful soffit reflections and diffusion surfaces in halls but can suffer from poor sight lines to the stage.		
CATT	Room acoustic prediction and auralization software. CATT is run by Bengt-Inge Dalenbäck (PhD) in Sweden.		
Catwalks	Catwalks are elevated narrow walkways used by theater personnel for technical functions such as lighting and sound.		
Cheek walls	Side walls of a hall within the throat wall zone that splay out to provide early reflections. Usually the first portion of the throat area.		
Clarity index (C80)	C80 is a measure of the strength of early reflections reaching the listener. It is defined as the logarithmic ratio between the sound energy reaching the listener in the first 80 milliseconds to the energy received after 80 milliseconds. 80 milliseconds is chosen as an indicator of clarity for music, whereas 50 milliseconds is typically used for speech clarity.	$C80 = 10\log_{10}\dfrac{\int_{0}^{0.08} p^2\,dt}{\int_{0.08}^{\infty} p^2\,dt}$	
Concert hall shaper	Type of innovative orchestra shell used to increase resonance around the symphony orchestra and front of the seating area.		

Term	*Definition*	*Equation*	*Variables*
Concrete masonry unit (CMU), concrete block, cinder block	A hollow masonry unit typically used in the construction of walls. High-density block made with sand is preferred over lightweight aggregates in multi-use halls. Partially grouted, and fully grouted walls use grout materials to fill the cores to increase mass.		
Cone of acceptance	Geometric angle of sound arriving at the listener where early reflections from side walls increase clarity. Reflections from outside this cone are considered not to improve clarity (such as from overhead or from below).		
Diffusion	Diffusion refers to the distribution of sound energy within a space. A diffuse space has an even spread of sound energy at all frequencies throughout, and the reverberation time (RT) would sound constant to a listener moving about. A nondiffuse space would have certain "hot spots" with highly focused sound energy, whereas other areas may sound dead.		
Dry or dead	Lacking in reverberation. A sense that the room is swallowing the sound, and there is no ringing or tail to the sound.		
Drywall	We use this term interchangeably with gypsum board or plaster board. A common building material in commercial buildings that is cheaper than plaster. Mounted on metal framing called studs using screws. Multiple layers reduce low-frequency absorption, and care must be used to avoid excessive absorption.		

Term	Definition	Equation	Variables
Early decay time (EDT)	EDT is defined as the first 10 dB decay time multiplied by 6. Therefore, it would be equivalent to the RT60 measurement if the decay is linear over 60 dB. However, often, the response is not linear, and studies have shown that a listener's perception of the liveness of a room is more closely related to EDT than RT.		
Early reflections	A term that encompasses both initial time delay gap (ITDG) and C80 reflections. Normally, the reflections that occur off surfaces that would reach the listener less than 80 milliseconds after the direct sound. Creates intimacy, clarity, immediacy, and increased loudness.		
Echo	A long-delayed sound reflection returning to the source. In this book, an echo is a negative factor because it creates difficulty for the performer. Often occurs with poorly shaped rear walls.		
Electronic architecture (EA)	An EA system can help enhance RT, intimacy, and clarity as well as improve the underbalcony experience for symphonic and opera productions. It uses microphones, loudspeakers, and sophisticated processing gear. Discussed in detail in Chapter 16.		
Envelopment	Envelopment is the feeling of being immersed in sound. It is similar to spaciousness but is affected by more than just the early lateral reflections. Envelopment is influenced by reflections from all directions (including overhead and behind) in the reverberant field. Envelopment is the sense that sound is arriving at the listener from all directions.		

Term	Definition	Equation	Variables
Eyring equation for RT	A more general form of Sabine's formula that offers a better prediction of RT in dead rooms. For rooms where adjustable absorption is used, Eyring is a better predictor of measured RT. Sabine underestimates the effect of high levels of sound absorption from banners and drapes in multi-use halls.	$$T_{60} = \frac{kV}{-S\ln(1-\bar{\alpha})}$$	T_{60}: reverberation time [s] k: constant [0.049 s/ft or 0.16 s/m] V: volume [ft³] or [m³] S: total surface area [ft²] or [m²] $\bar{\alpha}$: average absorption coefficient of all surfaces
Forestage reflector	A ceiling element akin to the orchestra ceiling but located over the stage extension area. Used to aid in onstage hearing and blending of classical music and pit reflections. Also useful in supporting speech with early sound reflections.		
Forestage reflector zone	Generally considered the zone in plan view, above the stage extension past the proscenium opening, used to locate the ceiling reflectors.		
Gallery seating	These are similar to box seats but tend to be larger sections of 8–12 seats on side walls that are accessed internally. Useful for soffit reflections.		
Glass fiber–reinforced gypsum (GFRG)	A thin plaster factory-shaped panel that can be made into complex forms for use on balcony faces and for other diffusive shapes. Used at Alice Tully Hall for high-frequency diffusion on side walls. Caution: avoid excessive LF absorption with thick materials or backplastering.		
Heating, ventilation, and air-conditioning (HVAC) systems	In this book, we are only concerned with the noise generated by these systems and its effect on the acoustics. Quieter is better. There are many excellent texts on this subject, and they are not covered in depth here.		

Term	Definition	Equation	Variables
Hertz (Hz)	The frequency of sound or cycles per second. The hearing range for normal listeners is about 16–15,000 Hz.		
High frequency	High frequency refers to sound between 4–15 kHz. 15 k is chosen as the max limit because it is the upper limit of human hearing for most people.		
Immediacy	How quickly a hall responds to a note being played on the stage. Created by early reflections around the musicians and in the throat area.		
Initial time delay gap (ITDG)	ITDG is the time difference between the direct sound and the first reflection that reaches the listener. It is a critical element in achieving clarity and presence in music and is usually determined from plans. Target values are less than 20 milliseconds, and intimacy suffers when they are beyond 35 milliseconds.		
Intensity (*I*)	Intensity is the time-averaged rate of energy transmission through a unit area normal to a specified direction, often the direction of wave propagation. In the equation, p is the acoustic pressure, u is the particle velocity, and T is the time period integrated over. For spherical free-progressive waves in the direction of propagation, the intensity is $I = \dfrac{P_e^2}{\rho_0 c}$, where $\rho_0{*}c$ is the characteristic impedance in a medium with density, ρ_0, and sound speed, c.	$I = \dfrac{1}{T}\displaystyle\int_0^T pu\,dt$	*I*: intensity [W/m^2] *T*: time period of integration [s] p: acoustic pressure [Pa] u: particle velocity [m/s] ρ_0: density of medium [kg/m^3] c: speed of sound in medium [m/s] P_e: effective pressure amplitude [Pa]

Term	Definition	Equation	Variables
Interaural cross-correlation (IACC)	IACC measures the difference between the sound reaching a listener's left and right ear. It is a tool that helps quantify diffusion, spaciousness, and envelopment. The measure of the difference in signals received by two ears of a person. IACC values range from −1 to +1. A value of −1 means that the signals are identical but completely out of phase. A value of +1 means that they are identical, and 0 means that they have no correlation at all. The IACC will be nearly +1 for mono sources directly in front of or behind the listener, with lower values if the source is off to one side. Research shows that large IACC values correspond to greater degrees of envelopment in multi-use halls.	$$\text{IACC} = \max \left| \frac{\int_{t_1}^{t_2} p_L(t) p_R(t+\tau)\, dt}{\left(\int_{t_1}^{t_2} p_L^2\, dt \int_{t_1}^{t_2} p_R^2\, dt \right)^{1/2}} \right|$$ for $-1\text{ ms} < \tau < +1\text{ ms}$	p_L: sound pressure at left ear [Pa] p_R: sound pressure at right ear [Pa] t_1 to t_2: time period of measurement, often 0–1000, 0–80, or 80–1000 milliseconds where time 0 is the arrival of the direct sound [s] τ: time range, typically varied over −1 to +1 ms [s]
Intimacy	Intimacy is feeling close and connected to musicians or singers on stage. Intimate halls have an enhanced connection between the audience and the performer. Highly prized in hall design. Smaller halls are usually more intimate, but many factors can impact intimacy including initial time delay gap, loudness, and early-to-late sound reflections.		
Lateral fraction (LF)	LF is a measure of the lateral reflections reaching a listener. It is measured by comparing the sound received by a figure-8 microphone with its null direction aimed at a source to a nondirectional microphone in the same location. In the equation, the figure-8 measurement begins at 5 milliseconds to make sure that the direct sound is eliminated. LF is closely correlated with source width, i.e., a high LF spreads and widens the sonic image of music, a pleasing effect.	$$\text{LF} = \frac{\int_{0.005}^{0.08} p_8^2\, dt}{\int_{0}^{0.08} p^2\, dt}$$	p_8: measurement from figure-8 microphone with null direction pointed at source

Term	Definition	Equation	Variables
Line sets	In order to raise and lower scenery, theatrical lighting, orchestra shell ceilings, and theatrical devices, pipes running left and right in the stage house are attached to cables that facilitate the movement of the line set.		
Loge	A theater loge is a private seating chamber. It can take on a variety of different forms. Some are completely enclosed rooms in which the show can be viewed through a window, and others are simply partitioned-off seating sections.		
Loudness (G) level	G level, also called total sound level or strength, is an indicator of loudness relative to a distance from a source. It is defined as the ratio of the direct sound response at a listener location to the response at 10 m from the same source in an anechoic chamber.	$$G = 10\log_{10}\frac{\int_0^\infty p^2\,dt}{\int_0^\infty p_A^2\,dt}$$	p_A: sound pressure at 10 m from source [Pa]
Low frequencies	Here, we use the definition of low frequency or bass interchangeably. It is usually sound energy from 125 Hz to the lower frequency limit of hearing, about 20 Hz.		
Mean free path (m)	In acoustics, the mean free path is the average distance that a sound wave travels between collisions with boundaries.		
Medium-density fiberboard (MDF)	A dense material used in halls for walls and acoustic reflectors and often used with wood veneer finishes. MDF has a typical density of 600–800 kg/m³ or 0.022–0.029 lbs/in³ and is denser than plywood or particle board, making it a good candidate for multi-use halls.		

Term	Definition	Equation	Variables
Noise criteria (NC)	NC is a set of curves to quantify the overall level of noise in a space. They are spectral sound pressure level (SPL) curves that are roughly based on equal loudness contours. The lowest NC curve, which is not exceeded by an SPL measurement at any frequency, would be the measured NC level for that space. Target NC-15 for most multi-use halls.		
Noise reduction coefficient (NRC)	Used to rate the sound absorption of common acoustic materials such as panels and ceilings. An average of coefficients in the mid-frequency octave bands (250–2000 Hz).		
Orchestra pit	An orchestra pit is the location of musicians for a ballet or musical theater performance. It is referred to as the pit because it is usually a lowered area between the audience seating and the front of the stage, often extending under the stage.		
Orchestra pit lift	An orchestra pit lift is a system used to raise and lower the orchestra pit platform. It can be raised to orchestra level for additional seating when there is no pit orchestra, or it can be raised to stage level and used as a stage extension.		
Orchestra shell	An orchestra shell consists of side and overhead acoustic reflectors on stage behind the proscenium. The shell is used to provide useful early reflections to musicians on stage, to provide a blending chamber for sound, and to project sound out to the audience. Also called the concert enclosure.		

Term	Definition	Equation	Variables
Parterre	Parterres are slightly raised separated seating zones at the rear and sometimes sides of the orchestra-level seating. They are used to divide large seating zones to reduce the scale of the large seating bank, increasing a sense of intimacy with the performers.		
Pit rail	The pit rail is a rail separating the pit opening from the audience seating. It can be in a number of possible positions and performs various acoustic functions.		
Plenum	Zone or volume used to deliver or collect air. It is usually located below the seating on various levels and is used for the quiet delivery of conditioned air up through openings in the floor.		
Presence	A sense that the listener is close to the musician. This occurs when there are strong early reflections in relation to the reverberation.		
Programming	Programming is the planning of the functional and operational requirements for the spaces within a building.		
Proscenium	The proscenium is the wall and opening that separates a stage house from the remainder of the hall. It serves as a sort of "window" through which the audience views the performers on stage.		
Psychoacoustic	Those factors other than scientific acoustic criteria that affect the perception of acoustics.		
Resonance	Sometimes the same as RT. A warm, rich resonant hall has a strong RT and BR. Resonance is also technically defined as an architectural system that is reinforcing certain driving frequencies, but that rarely occurs in large halls with significant dimensions discussed in this book.		

Term	Definition	Equation	Variables
Reverberant field	An area of the hall where the reflections dominate instead of the direct sound field. Typically the upper reaches of the hall far from the source.		
Reverberation time (RT)	RT measures how long it takes for a sound to decay to inaudibility in a space. It is defined as the length of time required for the SPL to fall by 60 dB after the source has stopped. A direct measurement of this 60-dB drop is referred to as the RT60. However, ambient sound often dominates before a sound decays by 60 dB. The RT30 or RT20 is a more common measurement. It is the time it takes for the sound to decay by 30 or 20 dB, respectively. These time measurements are then multiplied by 2 or 3, respectively, such that the RT30, RT20, and RT60 should be identical if the decay is completely linear. The Sabine equation is the most widely used method for predicting RT within auditoria. It relates the RT with the volume (*V*) and sound absorption (*A*) of the space. Sound absorption is frequency dependent and measured in "sabins" (ft²) or "metric sabins" (m²). It takes into account finishes, people, air conditions, and other factors that contribute to absorbing sound waves. "*k*" is a constant and is taken as either 0.049 seconds/ft or 0.16 seconds/m.	$RT = k\dfrac{V}{A}$	RT: reverberation time [s] V: volume [ft³] or [m³] A: sound absorption [sabins] or [metric sabins] k: constant [s/ft] or [s/m]
Rigging	In a multi-use hall, the stage house is equipped with devices that raise and lower scenery, lights, and acoustic shell ceilings. These can be either motorized (called winches) or manual counterweighted sets. The counterweight simply balances the weight of the moving set, making it easier to move manually.		

Term	Definition	Equation	Variables
Seat wagon	In order to easily move sections of seats on and off the pit lifts, seats can be affixed in large groups to a castored platform called a seat wagon. The wagons then store in areas below or off stage.		
Soffit	A soffit is the underside of a building element. Throughout this book, it often refers to the underside of a balcony, side box, or side gallery.		
Sound intensity level (IL)	IL is a way to express intensity in decibel units. It is 10 times the logarithm of the ratio of intensity to a reference intensity, I_{ref}, which is taken as 10^{-12} W/m² in air.	$IL = 20\log_{10}\dfrac{I}{I_{ref}}$	
Sound pressure level (SPL)	SPL is a way to express sound pressure in decibel units. It is 20 times the logarithm of the ratio of effective sound pressure amplitude to a reference sound pressure, P_{ref}, which is taken as 20 µPa in air. This is the most commonly used decibel quantifier of sound because it can be easily measured with a pressure microphone.	$SPL = 10\log_{10}\dfrac{P_e}{P_{ref}}$	
Sound-transparent ceiling	A ceiling in a hall that is visually closed but open sonically to allow free passage of uncolored sound.		
Spaciousness	Spaciousness is a term used to describe the size of the acoustic image created by the source. A source that lacks in spaciousness would appear to come from a single point stage, whereas a source that is full sounding appears to come from the full width of the stage, even beyond. The strength of lateral reflections, or reflections that reach listeners from the side, strongly influences spaciousness.		

Term	Definition	Equation	Variables
Specular	Specular reflections are mirror-like as opposed to diffuse or blended reflections.		
Stage house	A multi-use hall often has a large volume behind the proscenium opening that allows rigging to operate freely. This tall, wide, and deep space is called the stage house or sometimes the fly loft. In this book, the stage house houses the orchestra shell, allowing the hall to be used for a wide range of music performances.		
Tail	The reverberant tail refers to the late portion of the decay curve for a room. A smooth tail indicates that sound gently fades out within the space, whereas an echo would be indicated by a sudden energy spike in the tail.		
Theater consultant	A critical member of the design team who, along with the architect and acoustician, define the theater technology, audience sight lines, and backstage planning to make theatrical facilities function. This person is vital to detailing and specifying the shell, adjustable acoustics, stage extensions, and pit lifts that make a hall multipurpose.		
Throat	The throat of a hall is the side walls that typically splay out from the proscenium opening to the full width of the hall usually in the front one-third to one-half of a hall. See comments above on cheek walls.		
Timbre	The quality of sound that distinguishes one instrument or singer from another. It can be harsh or mellow, for example. A good hall does not change or distort an instrument's timbre.		

Term	Definition	Equation	Variables
Trap room	The trap room of a theater is a multi-use space beneath the stage. Primarily, it is used for access to the stage via trap doors, hence the name.		
Velour	Heavy, thick fabric sometimes incorrectly called velvet; used on stage for theatrical drapes. Often used as adjustable acoustic systems when doubled up in folds.		
Volume	Volume (*V*) is referred to in this book as the contained quantity of air in a space defined by walls, floor, and ceiling (or roof). Often, the volume is not clearly defined geometrically by the plans and sections.		
Warmth	Low-frequency reverberation between 75 and 350 Hz. Warm halls have higher levels of reverberation in this range. Closely correlated to BR.		
Wool serge	A natural fabric made of woven wool (30 oz.) that is durable and stable. Highly absorptive when used over a contained air space. Used in adjustable acoustic systems.		

References

Bate, A.E. and Pillow, M.E. *Proceedings of the Physical Society*, 59(4): 535–541, 1947.

Beranek, L. *Concert and Opera Halls: How They Sound*, Acoustical Society of America, 1996.

Beranek, L. *Concert Halls and Opera Houses: Music, Acoustics, and Architecture*, 2nd edition. Springer-Verlag, New York, 2003.

Beranek, L.L. *Noise Reduction*. Peninsula Pub., Los Altos, CA, 1991.

Bustard, C. Review: Richmond Symphony. The Virginia Classical Music Blog, September 26, 2009.

Bustard, C. Review: Richmond Symphony. The Virginia Classical Music Blog, February 28, 2010.

Cantrell, S. Dallas City Performance Hall proving a boon to classical-music groups. *The Dallas Morning News*. Availlable at http://artsblog.dallasnews.com/2013/09/dallas-city-performance-hall-proving-a-boon-to-classical-music-groups.html/ (accessed September 24, 2013), 2013.

Daegu Opera House Foundation. Acoustics. Available at http://english.daeguoperahouse.org/facili/sub2_2.asp (accessed September 1, 2015).

Daniel, D. The 10 Greatest Opera Houses. *Travel + Leisure Magazine*, March 1999.

Diller, E., Scofidio, R., and Renfro, C. *Diller, Scofidio + Renfro: Lincoln Center Inside Out: An Architectural Account*. Damiani, Bologna, Italy, 2013.

Dixon, G. In the Limelight—MOSC at Wagner Noël Performing Arts Center, November 15, 2011. Available at http://www.inthelimelightonline.com/?p=593.

Gay, W.L. Douglas Perfects his Balancing Act. *Fort Worth Star-Telegram*, February 20, 1999.

Gelfand, J. (Reporter). Best of 2003: Classical Music. *The Cincinnati Enquirer*, Sunday, December 28, 2003.

Goldsmith, M. *Discord: The Story of Noise*, reprint edition. Oxford, UK, 2014.

Knudsen, V.O. and Harris, C.M. *Acoustical Designing in Architecture*, 1st edition. John Wiley & Sons, Inc., New York, 1950.

Knudsen, V.O. and Harris, C.M. *Acoustical Designing in Architecture* (originally published in 1950), John Wiley & Sons, Inc., New York, 1980.

Midland Reporter-Telegram. Wagner Noël Raises Standard for Arts, Entertainment in the Basin, November 3, 2014. Available at http://www.mrt.com/top_stories/article_484178e0-6374-11e4-b098-630f834624d3.html.

Nancy Lee and Perry R. Bass Performance Hall. The Hall. Available at http://www.basshall.com/thehall.jsp (accessed September 1, 2015), n.d.

Rettinger, M. *Acoustic Design and Noise Control*, Volume II. Chemical Pub., New York, 1977.

Sabine, W.C. Architectural Acoustics. *Journal of the Franklin Institute* CLXXIX(1): 1–20, 1915. Available at http://www.sciencedirect.com/science/article/pii/S0016003215907962 (accessed September 24, 2015).

Schultz, T.J. *Acoustical Uses For Perforated Metals: Principles and Applications*. Industrial Perforators Association, 1986.

Swed, M. Music Review: St. Lawrence String Quartet Brings Edge to Wallis. *Los Angeles Times*, January 16, 2014. Available at http://www.latimes.com/entertainment/arts/culture/la-et-cm-lawrence-wallis-review-2-20140117-story.html (accessed September 24, 2015).

Tarradell, M. Sarah Jaffe and the Triple Play brilliantly break in new Dallas City Performance Hall with robust concerts. Pop Culture Blog, September 15, 2012. Available at http://popcultureblog.dallasnews .com/2012/09/sarah-jaffe-and-the-triple-play-brilliantly-break-in-new-dallas-city-performance-hall -with-robust-concerts.html/ (accessed September 24, 2015).

Vankin, D. Wallis Annenberg Center for the Performing Arts Set for Grand Opening. *Los Angeles Times*, October 17, 2013. Available at http://www.latimes.com/entertainment/arts/culture/la-et-cm-wallis -annenberg-center-sneak-peek20131017-story.html (accessed September 24, 2015).

Wakin, D.J. Musicians Hear Heaven in Tully Hall's New Sound. *The New York Times*, January 28, 2009. Available at http://www.nytimes.com/2009/01/29/arts/music/29tull.html?_r=3&th (accessed September 24, 2015).

WTVR. Richmond CenterStage Opens This Weekend. An Interview with WTVR, September 9, 2009.

Index

Page numbers followed by f indicates figures.

Milton Keynes UK
Ingram Content Group UK Ltd.
UKHW050309111024
449327UK00049B/379

9 780367 866105